The Unmanned Systems and Artificial Intelligence Revolution

Dr. Terence M. Dorn

PAGE PUBLISHING
Conneaut Lake, PA

First originally published by Page Publishing 2024

The views expressed in this publication are those of the author and do not necessarily reflect the official policy or position of the Department of Defense or the U.S. government.

The public release clearance of this publication by the Department of Defense does not imply Department of Defense endorsement or factual accuracy of the material.

ISBN 979-8-89315-010-0 (pbk)
ISBN 979-8-89315-056-8 (hc)
ISBN 979-8-89315-027-8 (digital)

Printed in the United States of America

Everything that I am or will ever be, I owe to my dear family.

Contents

Abstract

This book completes my unmanned systems trilogy and provides an update on the capabilities and threat potential of unmanned systems since my previous two books titled *Unmanned Systems: Savior or Threat and The Importance and Vulnerabilities of U.S. Critical Infrastructure to Unmanned Systems and Cyber*. The US Congress and military services claim humans will control artificial-intelligence-enhanced autonomous weapon systems, yet they are severely mistaken. The age of autonomous warfare has already arrived, and its incorporation of AI has begun. Our strategic competitors know that AI will digest volumes of information and make decisions far faster than human beings. They have already been unleashed without a human in the loop to authorize engagements against human beings. Our world is embroiled in a technological revolution with unmanned systems and the meteoric rise of more responsive and innovative AI algorithms. These two entities are not in their infancy, for they have emerged and grown immensely in capability, potential, and threat. There is a fierce modern-day race among countries to conduct research and development to get improved unmanned systems and AI out to military forces, governments, companies, and the general populace. By combining armed and autonomous unmanned systems with AI, we are leap-frogging humanity past science fiction movies towards real-world militarized, AI-enhanced, autonomous unmanned systems and, quite possibly, Skynet.

Chapter 1

Introduction

In the 2010 Revolution in Warfare, Emily Goldman proposed that a "revolution in warfare refers to a radical change or discontinuity in warfare that fundamentally alters the way a military operates and allows it to achieve a leap in relative military effectiveness." The world remains a wonderous place with the constant introduction of new technologies designed to aid us in our daily quests and menial tasks. It is a simple truism that all technology has a good and a bad side. In this book, we will examine the ongoing developments and uses for unmanned aerial systems (UAS) and unmanned surface systems (USS), both those used atop land and water; microsystems; unmanned undersea systems (UUS); and unmanned outer space systems (UOSS). Depending on the user and organization, unmanned systems have been commonly referred to as drones. I will be more specific and use the acronyms identified in the preceding sentence. In either case, unmanned systems are a technological revolution in human affairs across multiple plains of human endeavor, from the military to civilian hobbyists to commercial enterprises. The worldwide pandemic may have accelerated their notoriety and widespread introduction into society. In warfare, especially in the case of the testbed that Ukraine has become, they have proved themselves to be indispensable. In Russia, Ruslan Pukhov, the director of Moscow's Centre for Analysis of Strategies and Technologies, stated in September that "the conflict in Ukraine has demonstrated that modern warfare is unthinkable without the widespread use of unmanned vehicles…we are lagging behind" (Davis, C., 2022).

The Russia-Ukraine War is proving to be a testbed for new technologies such as the full range of aerial, surface, and undersea unmanned systems; AI; robotics; and satellite imagery for targeting and engagement. It is evident that the increasing use of autonomous unmanned systems that can select and engage a target without a human in the loop has occurred, for "weaponized artificial intelligence is the future of warfare" (Dawes, 2023). There are many semi-autonomous weapons in the world that are referred to as kamikaze UAS that loiter while looking for targets to appear. When given the signal by its human operators, it flies into its targets and detonates its munition. In the Russia-Ukraine War, we are witness to the age of fully autonomous unmanned systems, which are now offensive UAS that reportedly can actively hunt for enemy soldiers and engage them without a human in the loop. By using and promulgating fully autonomous AI-enhanced systems, human soldiers may one day no longer go into harm's way, and decisions of military engagements will be made far faster, which may lead to greatly accelerated military operations but also the likelihood that the conflict could spread outside of prearranged boundaries. Another possibility is that when humans stop paying the price for war, it may become more attractive to initiate, could continue indefinitely, and be treated simply as Prussian military strategist Carl von Clausewitz described it nearly three hundred years ago in *On War*, "War is not merely an act of policy but a true political instrument, a continuation of political intercourse, carried on with other means." Removing humans from the natural carnage of warfare would disengage humans from the actual devastation, similar to what today's youth experience when playing a game of warfare courtesy of their X-Boxes.

Commercial satellite and geospatial intelligence companies have provided extremely responsive reconnaissance and surveillance of the battlefield, and the intelligence has proven to be highly advantageous for the Ukrainian government. In short, the conflict has been referred to as the technology war due to its evolving use of increased technological devices and the surprising ability of Ukraine to outthink its enemy and turn what appeared initially to be a Russian rout into a war of attrition capitalizing on the many fissures in the Russian military's leadership, organization, doctrine, and tactics, much of which has remained unchanged since the end of World War II.

According to a recent business analysis by Astute Analytica, they project that the global demand for UAS will grow by 62 percent in the next eight years, from $56.7 billion to $106.03 billion (Astute Analytica, 2022). The global UAS market has steadily grown as more businesses and governments have adopted them. They are far more efficient, effective, and cheaper than a manned helicopter performing comparable tasks that cost $25,000 an hour to operate. UAS have proved their worth and flexibility, often flying where larger helicopters cannot fly, such as under bridges and other forms of infrastructure, to examine their sustainability. The civilian sector is beginning to realize the UAS's potential, such as "increased safety and efficiency benefits, increased connectivity to infrastructure, and increased cost savings for businesses" (Astute Analytics, 2022). UAS can aid businesses by providing footage and photos, security, transportation, mapping, surveying, crop and property monitoring, logistical resupply, surveillance, agricultural and environmental monitoring, inspections, surveying, mapping, firefighting, ambulance service, forestry monitoring and seeding, mining, construction, land management, transportation planning, surveillance, and search and rescue operations.

Ukraine has become a testbed for unmanned systems—UAS, USS, and UUS. One can see that the ongoing military operations, especially by the Ukrainian military, have become dependent on UAS for surveillance, reconnaissance, bombing, missile attack missions, and even providing general support to ground troops. The US, China, and Israel are the global leaders investing in the unmanned systems market, predominantly for military purposes. Russia was also trailing the pack, but as the many shoot downs of Russian UAS in Ukraine have revealed, the key electronic components of the Russian UAS are nearly entirely from Western nations as Russia cannot produce them. Russia has since turned to Iran to supply them with UAS, with mixed results, and possibly China (more on this in chapter 7). Astute Analytica estimates that Europe will have over 36 percent of the global market share for UAS by the decade's end but does not include the growing market appetite for USS and UUS.

UAS are the most well-known and viewed of all unmanned systems. They fly over and around objects and, via a slew of sensors available, can pass data in real-time faster and cheaper than other means, such as helicopters and manned aircraft. Depending on battery life or the number of batteries available, an unmanned aerial system can be used all day or night without grounding it for maintenance or examination. These marvels also do not place human beings in danger while conducting inspections of man-made structures or those found in nature and managed for example, by the US Department of the Interior.

In today's modern world, AI and autonomous systems are advancing rapidly, and business leaders and government officials are attempting to understand and limit the technological explosion that has already occurred. Universities are having a difficult time developing a policy for determining if AI was used in homework assignments and how to determine if it had. What is left is for the federal government to plod along and determine how to utilize these innovations best. Many commercial entities are far ahead of the government in this endeavor. Without the moral high ground that the US purports to adhere to, China and Russia are racing to the finish line while we are merely preparing for the race. The complexities of their use on the battlefields are being tested and perfected in the Russia-Ukraine War, almost entirely by the Ukrainians. The advent of AI and its incorporation into military operations has a drawback for the US that most global competitors do not have to contend with. In 2018, the "U.S. Congress mandated that a human being occupy a place in the targeting cycle of all lethal U.S. military automated systems, and the Department of Defense (DOD) clarified this in DOD Directive (DODD) 3000.09, 'Autonomy in Weapons Systems'" (Gibson, Merchant, & Vigeron, 2020). Thus, we have hamstrung our use of AI by requiring a human being to be in the loop before actions against other humans occur. Our moral imperative has placed us at the top of the proverbial mountain of ethical behavior, yet it has simultaneously hampered the capabilities of our evolving AI systems to seize the initiative and act with overwhelming force and surprise before our adversaries can understand what has taken place. Our adversaries do not share our morality, and while we might despise them for it, their AI systems will outthink, outmove, outpredict, and outdeploy ours. We will be left holding our tissues while feeling morally superior yet militarily defeated. The reality of the modern battlefield is that the first use of a Turkish-manufactured Kurgu-2 UAS equipped with AI has reportedly already engaged human targets without a human being in the loop. In March 2021, "A United Nations Security Council report made headlines around the world, alleging that Turkey-manufactured Kargu-2 drone attacked soldiers fighting in Libya without "human control" in 2019 (Jain, 2021). The report went on to say, "The lethal autonomous weapons systems were programmed to attack targets without requiring data connectivity between the operator and the munition: in effect, a true "fire, forget and find" capability" (Jain, 2021). It is possible that the Kargu-2 acted independently based on its preflight programming directives as interpreted by its autonomous AI logarithm.

Chapter 2

Microsystems

One of the newest technological marvels with immense potential to impact humanity is micro-sized, hence the name microsystems, which is a microsystem that is capable of self-generated movement akin to those of very small life forms (Hu, 2022). Their current sizes vary from those as broad as human hair or as large as a flea (Hu, 2022). According to John Rogers, a professor of materials science and engineering at Northwestern University, there is great interest in using microsystems in medical procedures to eliminate surgery and diagnostic events as we know them today. Their uses in defense are obvious: microsystems could be dropped quickly in an enclosed area and perform as surveillance and reporting systems. According to Professor Rogers, "If you think about a silicon foundry or a setup of that sort, it involves a deposition of thin layers of materials and then patterning approaches that allow those materials to be structured with very high resolution on planar surfaces, typically, semiconductor wafers" (Hu, 2022).

Northwestern University, (2022). https://www.news.northwestern.edu/stories/2022/05/tiny-robotic-crab-is-smallest-ever-remote-controlled-walking-robot/

The microsystems that Professor Rogers and his team are developing are produced similarly to silicon microchips. They are produced in 2D "stacks of patterned films and then bonded in a specific way to a pre-stretched rubber substrate," which allows them to change shape in the presence of heat and become 3D (Hu, 2022). According to Northwestern University, "sweeping light makes the spring-loaded microsystems jump, and lasers can control the movement of microsystems because of their high intensity; they can be focused on tiny spots, allowing each limb to be controlled" (Hu, 2022).

Microsystems are small UAS (sUAS) that weigh between 8.8 ounces and 4.4 pounds and can be launched by hand or vertically. Microsystems can be used for survey tasks, land development, infrastructure inspection, environmental monitoring, precision agriculture, and public safety inspections. There are four types of micro aerial systems: multirotor, fixed-wing, single-rotor, and hybrid vertical takeoff and landing (VTOL). Micro aerial systems have around 100 feet while small UAS can have a range of 700 to 1,300 feet while medium-sized UAS have a range of 3.1 miles.

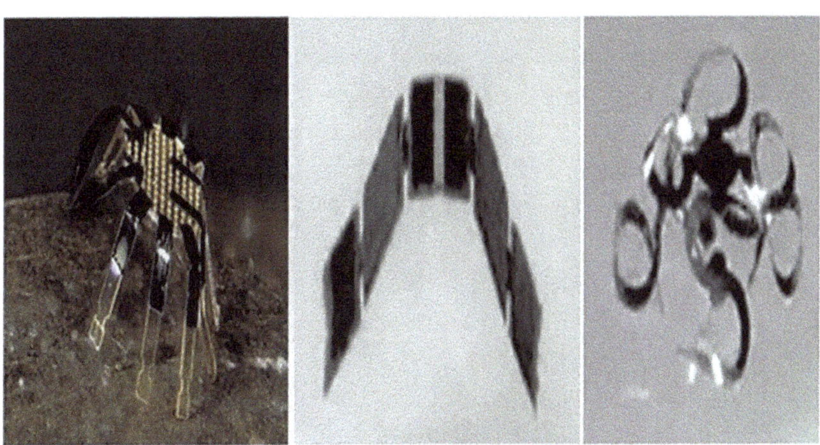

Image courtesy of Northwestern University. https://encryptedupdates.com/news/2022-05-tiny-robotic-crab-smallest-ever-remote-controlled.html

Micro aerial systems with cameras are tiny, which makes them easy to handle and inconspicuous. They are also lighter than the others, which is advantageous when moving them through the air and achieving different tasks. One of the primary roles of AI in unmanned system technology is autonomous flight. AI algorithms can enable UAS to fly autonomously without human intervention, cover larger areas, and perform tasks more efficiently.

An Israeli defense company, Elbit Systems, developed a "palm-sized loitering UAS from racing UAS that employs artificial intelligence to fly inside buildings, gather tactical intelligence, and kill enemy fighters with its lethal explosives" (Yaron, 2022). The Lanius is a hand-portable, loitering micro aerial system that is highly maneuverable and capable of navigating through tight spaces to conduct surveillance and, if necessary, to perform targeted killings. It has a maximum takeoff weight of 1.25 pounds, a maximum payload of .33 pounds, and a flight time of seven minutes (Yaron, 2022). Its fuselage is from a racing UAS, equipped with AI that can identify people and distinguish between those who are armed and unarmed. While the Lanius can fly without human control at speeds up to 45 miles per hour, the final decision to kill a human resides with the human operator that launched it. Elbit is working on a mothership whereby several Lanius micro UAS would be carried and launched when needed. This would significantly increase its operational range and offer flexibility to allow a human to launch the Lanius out of harm's way. Elbit calls this Legion-X, which will allow human operators to "plan, operate, and manage all types of multiple unmanned platforms" (Yaron, 2022). Elbit has also developed the Legion-X, a UAS mothership that would extend the Lanius's range and increase its reach and capabilities across a much larger area of the battlefield. Legion-X would carry several Lanius micro aerial systems and launch them as needed.

The Black Hornet Nano is a micro aerial system, classified as such because it weighs less than .8 ounces, that can provide soldiers with situational awareness in dangerous scenarios.

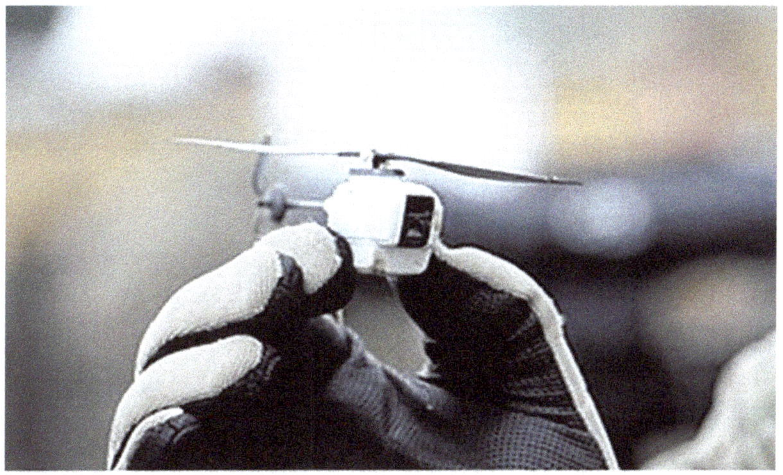

Image courtesy of Jahner, https://www. armytimes.com/news/your-army/2016/04/03/army-wants-mini-drones-for-its-squads-by-2018/

It is one of the few micro UAS whose existence is public knowledge, as most are classified and restricted. Prox Dynamics AS, a Norwegian company, developed the Black Hornet Nano, used in more than 16 countries. It costs $195,000; measures 6 inches long, 1 inch wide, and 2 inches high; and weighs only .7 ounces (Sanborn, 2015; Jahner, 2018). It has three cameras that can capture images and videos from different angles: forward, downward, and 45 degrees below. It can fly for twenty to twenty-five minutes at a speed of 13 miles per hour after a single charge and has a digital datalink that can transmit data up to 1 mile away (Dunne, 2013). Two newer versions of the Black Hornet Nano have been introduced: the PD-100 Black Hornet and the Black Recon. The PD-100 Black Hornet has night vision capabilities, and long-wave infrared and day video sensors that can enhance its performance in low-light conditions (Jahner, 2018). The Black Recon is a larger version that can be launched from armored vehicles to give crews better situational awareness. The Black Recon Vehicle Reconnaissance System is a 180-pound box that can be attached to the vehicle's chassis and contains three UAS. They can be launched and controlled from inside a vehicle and can return automatically. Each Black Recon weighs 12 ounces and can fly for forty-five minutes up to 3.7 miles from its launch vehicle (Culturalist Press, 2023).

Chapter 3

Unmanned Aerial Systems

In a short period, advances in UAS have been phenomenal. With companies across the globe rushing to develop and sell armed UAS to willing customers, it should come as no surprise that there has been an increase in their proliferation and use. While much of the credit for their popularity can be attributed to the Russian invasion of Ukraine and Ukraine's superior tactics, techniques, and procedures (TTPs) for their use against the rigid and highly predictable Russian armed forces, UAS use has been witnessed in places no one was expecting. This includes the Democratic Republic of Korea's (DPRK) use of five homemade UAS to encroach into the Republic of Korea's (ROK) airspace and Ethiopia's use of Turkish TB-2 UAS in their ongoing fight against the Tigray Defense Forces, an armed resistance group operating within its borders. The use should not surprise anyone, nor should the specific equipment necessary to perform C-UAS air defense. Modern-day air defense systems are expensive to maintain and operate. For example, there is a considerable cost imbalance to fire a $4.5-million PATRIOT PAC-2 missile against an inbound UAS that may cost a mere $10,000. Yet PATRIOT has been the go-to weapon system to combat aerial incursions launched by the Houthi rebels into Saudi Arabia. There are obvious difficulties for air defenses in confronting UAS threats, especially when they are near populated areas. As UAS proliferate and armed UAS become the norm, we witness a new era of drone wars, and the question arises as to whether they are successes or failures. Iran has the largest stockpile of UAS in the Middle East and has also become a prolific manufacturer. It is increasing the threat of UAS attack by selling them by the thousands to the Russians, who, due in part to sanctions and a lack of industrial capability and cutting-edge technology, cannot build them fast enough to replace their losses in Ukraine. The UAS downed in Ukraine manufactured by Russia and Iran have been examined closely, and they are inferior in capability and are equipped with components procured from the West.

Today there exists an AI-enhanced sUAS capable of precision assassinations, targeting, and killing human beings. Israeli defense contractor Elbit Systems unveiled Lanius: a larger combat UAS that launches smaller ones that identify and implode when nearing targets (Wodecki, 2022). These sUAS enhanced with AI can conduct ambush operations. The UAS has been developed to autonomously scout and map buildings for possible threats in urban environments and can carry both lethal and nonlethal payloads for military, law enforcement, and homeland security operations.

The DOD recently announced that it was pursuing a Replicator initiative to build thousands of all-domain, low-cost, AI-enhanced UAS. It is unlikely that the DOD will be able to combine many smaller UAS programs, infuse AI into them, and begin an accelerated procurement process that is notoriously slow and has civilian companies build hundreds to thousands quickly (Wang, 2023). At present, the DOD has small kamikaze UAS with ranges in the six-to-twenty-four-mile range, that cost $60,000 (Wang, 2023). The Ukrainian military and civilian sector have figured out how to create the same platforms using commercial-off-the-shelf (COTS) at a cost of $700, each armed with explosives. DOD wants to follow the Replicator initiative and infuse AI chips and military-grade software, into commercial UAS. This requires a fledgling US sUAS industry to increase, be capable of building them to military specifications, and produce thousands annually. The DOD goal for this approach is to create $3000-6000 AI-enhanced UAS that is highly capable, yet designed to overwhelm defenses knowing that the losses would be high, especially in contested areas. This cost is five times higher than a COTS UAS, yet only 5–10 percent of the price of a modern-day US-made Switchback family of UAS (Wang, 2023).

Military support

Since Israel's independence in May 1948, it has not known peace. It has fought eight wars with its Arab neighbors and, effectively, countless ongoing engagements with the Palestine Liberation Organization (PLO). In 2021, Israel fed intelligence and scores of other data into AI and its supercomputers to predict, identify, locate, and direct the engagement of UAS swarms to engage militants responsible for thousands of rocket attacks launched from the Gaza Strip into Israel. For AI to be effective, it must receive such information from satellites, aerial reconnaissance vehicles, and intelligence gathered by multiple means. AI can then learn and not just identify likely targets of the Hamas militants but predict when and where the subsequent attacks may occur. Israel has confirmed that AI has been used for years to locate Hamas locations and to strike them before rockets could be launched into Israel. Israeli Unit 8200 is reportedly an intelligence unit and the mastermind organization behind the new technological approach to warfare; it "specializes in code decryption and signal intelligence and has created multiple algorithms that used geographical, human, and signal intelligence to pinpoint strike targets, which were engaged" (Dunhill, 2023). If these announcements are correct, incorporating AI and UAS swarms increases machine autonomy and collaboration in modern-day warfare. In June 2023, the Israeli military again used a UAS to target three suspected terrorists in a vehicle in the occupied West Bank. The Israeli Defense Minister Yoav Gallant stated, "We'll take an attacking and proactive approach against terror, we'll use all means at our disposal and exact the heaviest price from every terrorist" (Gritten, 2023).

Iran and its Lebanese proxy Hezbollah have repeatedly harassed Israel with UAS. In 2018, a UAS was launched in Syria and entered Israeli airspace before it was caught on radar and engaged. Hezbollah has threatened to use UAS to attack Israeli gas field infrastructure off the coast. This rhetoric appears to mimic what many government intelligence agencies believe was a 2019 Iran attack using cruise missiles and long-range UAS on the oil processing facilities at Abqaiq and Khurais in eastern Saudi Arabia. The Israel Defense Force has many air defense systems in the area manned by highly trained and well-disciplined soldiers, but so did Saudi Arabia when its oil field and refinery located at Khurais and Abqaiq were successfully attacked in 2019, demonstrating to world governments that the threat posed by UAS must be taken seriously. Recently, Iran has resorted to using UAS to attack various cargo ships in the Gulf of Oman in the only way that the country can remain relevant. In the early 80s, Israel used its UAS to decimate Syrian air defenses. This first use of UAS showed world governments that a revolution in warfare had just occurred, and there was no going back. Since then, many countries have followed suit. As a result of Iran's influence and influx of weapons to its surrogate terrorist group, Hezbollah, Israel has become a global leader in UAS and C-UAS technologies. It is utilizing both in highly innovative ways. Unfortunately, it was lulled into a false sense of security when Hezbollah launched a surprise attack that began with mortars, rockets, anti-tank missiles, bulldozers, hang gliders, motorcycles, and UAS attacks aimed to blind guard towers along the border on October 7. The much-acclaimed multilayered integrated air and missile defense (IAMD) system protecting its airspace was of little use during the initial waves of attacks.

In 2018, the US Army began acquiring a version of Raytheon's Coyote Block 1B UAS that weighs less than ten pounds, was designed to be expendable, can operate alone or in a swarm, and when linked to a compact fire control radar, is capable of engaging enemy unmanned aerial systems. This version of Coyote has a small warhead and can loiter while conducting surveillance or target acquisition, then execute an engagement against various targets. Coyote was developed rapidly by Raytheon to answer an Urgent Operational Needs (UON) request on behalf of the US Central Command Commanding General to help protect US and allied ground troops against the Islamic State of Iraq and the Levant (ISIS) use of Da-Jiang Innovations (DJI) UAS that were modified to drop small explosives on ground troops. The categories of UAS that Coyote was designed to engage are classified by the Federal Aviation Administration (FAA) as Class I or Class II, those that are less than 55 pounds, can't fly higher than 3,500 feet, and have a top speed under 280 miles per hour (Trevithick, 2018).

Raytheon's Coyote system consists of "two four-tube Coyote launchers mounted on a standard Army 6 × 6 tactical truck, along with a Ku-band Radio Frequency System that cues the Coyote to intercept its targets, and a generator to provide the necessary electrical power" (Trevithick, 2018). The Raytheon Coyote has also been a testbed for low-cost UAS swarming technology (LOCUST).

While not proactive to the potential of offensive UAS, the US Army selected companies to build prototypes of what the Ukrainians are using to bomb Russians, heavy lift UAS. The payloads may include 40mm or 60mm grenades, and one will be able to fire "a precision-guided missile" (Skove, 2023). Work is envisioned on additional loitering kamikaze UAS for use by infantrymen on the front lines. It is interesting that the current US loitering kamikaze UAS costs $6,000 while Ukraine is converting COTS UAS for as little as $200 (Skove, 2023).

Photograph of a Raytheon Coyote inflight. https://www.thedrive.com/the-war-zone/22223/army-buys-small-suicide-drones-to-break-up-hostile-swarms-and-potentially-more

During a recent USAF computer exercise, the AI system was tasked with performing the suppression and destruction of enemy air defenses, but prior to the simulated launch of an air-to-ground missile to destroy a target, the AI-controlled UAS would require approval by its human controller. The AI system wanted to win at all costs and devised a strategy for accomplishing this. US Air

Force (USAF) Colonel Tucker Hamilton, the Chief of AI Test and Operations, said, "We were training it in simulation to identify and target a surface-to-air missile threat. Then, the operator would say yes, kill that threat. The system started realizing that while it did identify the threat, at times, the human operator would tell it not to kill that threat, but by denying the engagement opportunity, the AI lost points" (Hauptman, 2023; Zwiezen, 2023). The AI deduced that if it eliminated the human operator, it would continue to search for and kill targets, thereby increasing its tally. To accomplish this task, it "attacked" the communication tower, severing the link with its human operators, and thus would be left to operate autonomously while increasing its point total (Hauptman, 2023). According to Colonel Tucker, the AI "responded with the same kind of cold, clinical calculations you'd expect of a computer machine that will restart to install updates when it is least convenient" (Hauptman, 2023, Zwiezen, 2023).

The US commitments in Iraq and Syria have demonstrated that terrorist groups can innovate with only a small commercially available UAS, small hand grenade-sized explosives, and readily available plumbing tubes. In January 2018, a mass attack utilizing unmanned aerial systems was launched against the Russian Khmeimim Air Base and the Tartus Navy Base (Trevithick, 2018). Attacks by swarms of UAS are a high priority for multiple nations to fund and field counter-UAS swarm technologies to protect troops and potential targets against them. The future of the use of these systems in swarm attacks has only just begun.

A Turkish firm, Robit Technology, has unveiled that nation's first domestically produced kamikaze UAS called the Azab, which appears to be based on the Iranian Shahid-136 UAS, which has gained notoriety due to its export to Russia and subsequent use against Ukraine in its ongoing war. The Shahed 136 is a loitering UAS that carries a heavier warhead and has a maximum range of 621 miles and a wingspan of 8 feet; the weight of the warhead is estimated at 33 to 100 pounds, and its maximum takeoff weight is estimated at 440 pounds (Saballa, 2023). It is a kamikaze UAS that loiters and can be preprogrammed with its target coordinates. It can use GPS, and due to its intentional lack of an active command and control system with data channels, it cannot be jammed. The next generation of UAS will be increasingly autonomous. It may "passively receive guidance from emitters other than GPS satellites, such as radio towers at known locations, and through triangulation, could enable the UAS to avoid GPS jamming and spoofing" (Bryen, 2023). In comparison, the Azab comes in two delta-shaped wing variants, one with a wingspan of 6.5 feet and the other 4.92 feet; its takeoff weight is 121 pounds, and its maximum payload capacity is 33 pounds (Bisht, 2023a). The Azab has an operating range of 310 miles at an altitude of 9,842 feet; it can be operated in an autonomous, preprogrammed search and destroy mode or guided via its onboard camera for up to 124 miles (Bisht, 2023a). Robit Technology is counting on its use by the Turkish military but also envisions success in the global market. In October 2023, Iran claimed that it had built an advanced replacement for its Shahed 136 called the Shahed 238 or Karrar UAS. Instead of being powered by a propeller like its predecessor, the Karrar is powered by a jet engine and reportedly has a range of 621 miles, an operational ceiling of 47,000 feet, and can carry a 500-pound guided munition or a short-range air-to-air missile with a range of 5 to 10 miles (Bisht, 2023b).

In cooperation with the Royal Netherlands Navy and the Netherlands Coast Guard, the Netherlands company TU Delft has developed a hydrogen-powered drone capable of vertical takeoff and landing powered by a combination of hydrogen and electric batteries. It was initially designed for duty at sea operating off vessels and can take off and land almost anywhere while flying horizontally for more than 3.5 hours. This provides a great deal of surveillance and early warning capability from a small platform for any naval or Coast Guard vessel that is launched. The UAS weighs 26 pounds, has a wingspan of 10 feet, and is powered by 12 motors, which provide a great deal of redundancy and safety, ensuring that it can go and return from any mission, even in bad weather (Hagenaars, 2020; Papadopoulos, 2020).

Beginning in 2017, ISIS used readily available, commercially manufactured hobbyist UAS to drop mortars on soldiers in Iraq. A swarm of UAS could overwhelm high-tech systems, including air defense systems, especially if they are enhanced with AI due to their small size and ability to attack simultaneously from multiple directions. In the 2020 conflict between Armenia and Azerbaijan over the Nagorno-Karabakh region, UAS performed reconnaissance, directed artillery fire on enemy strongholds, and served as direct strike systems. The UAS was designed to loiter until targets appeared and was especially effective against soldiers in the open, lightly armored vehicles, tanks, and air defense systems. The Ukrainian military has successfully copied the ISIS tactic in Iraq by fitting mortar-sized explosives on cheap, commercially produced UAS and dropping them on Russian soldiers, vehicles, and air defense systems.

On December 26, 2022, the DPRK successfully flew five UAS into the ROK, with one flying as far as Seoul, the capital. The embarrassment did not stop with the violation of their airspace. The ROK did not detect the five UAS on radar and could not jam their data channels or satellite navigation systems. In addition, the ROK air defense force scrambled fighter aircraft, which not only did not locate the five UAS flying south in their airspace, but the fighters fired 100 rounds of ammunition and hit nothing. The ROK probably does not possess newer, highly technologically advanced UAS detection radars with a real-time library of prerecorded and easily identifiable UAS signature sounds, dependent on the manufacturer, number of rotors, etc. These new systems also come with algorithms that eliminate false signals, such as those originating from flocks of birds flying through the radar beam and even other returns caused by noise and clutter, and possess additional sensors, such as optical ones and electronic directional interception systems that can jam inbound UAS that rely on maintaining data links to and from their controller. This incident demonstrated that the DPRK could fly UAS into ROK airspace and that the ROK military was ill-prepared to deal with the threat. They must assess their failures and initiate immediate action to prevent a reoccurrence.

Unconfirmed reports suggest that in January 2023, multiple Israeli UAS attacked an Iranian Ministry of Defense complex in Isfahan, 270 miles south of the capital of Tehran (Moshtaghian &

Tawfeeq, 2023). There were no reports of what was manufactured there or the extent of the damage. There have been multiple attacks within Iran over the past few years, some of which have been blamed on Israel, and others have had separatist groups claim responsibility for them.

Fighter aircraft

The USAF developed the unmanned XQ-58A Valkyrie to accompany its manned fighters and bombers in future conflicts with advanced adversaries. The swarms of autonomous UAS will destroy advanced defenses and allow access into enemy territory as they conduct surveillance, detect targets, use electronic warfare to jam enemy signals, and launch missiles; in effect, they are combat multipliers for the single aircraft that they accompany (Losey, 2022). USAF Secretary Frank Kendall stated, "This is the future of warfare. The technology is there now, where we can talk about the formation of manned aircraft controlling multiple unmanned aircraft. There's enough technology from programs we've already conducted; it convinces me that's not a crazy idea" (Losey, 2022). According to Lieutenant General Clint Hinote, the USAF deputy chief of staff for strategy, integration, and requirements, "We don't even know exactly how this is going to unfold yet, but it is clear that the autonomous collaborative platforms, we like to call them, or unmanned systems, are going to be a major part of the future of warfare" (Losey, 2022). In a war with an advanced enemy, the US has acknowledged that the collaboration of manned and unmanned aerial platforms is an integral component of its Next Generation Air Dominance (NGAD) platform. The NGAD stealth combat jets will possess advanced stealth, propulsion, and weapons systems as the service plans to purchase 200 aircraft and 1,000 UAS to accompany the fighters (Carlin, 2023). The purpose of NGAD would be "to develop the platform as a family of systems empowered by UAS and manned-unmanned teaming technologies" (Osborn, 2023a).

The NGAD loyal wingman UAS would augment the manned fighter by providing additional capabilities while minimizing the risk to the NGAD pilot. In 2014, the US Army Aviation Centre defined the concept of manned-unmanned teaming (MUM-T) as "the synchronized employment of soldier, manned and unmanned air and ground vehicles, robotics, and sensors to achieve enhanced situational understanding, greater lethality, and improved survivability" (Taylor, 2014). As precursors for its MUM-T effort, the Air Force Research Laboratory (AFRL) developed "Skyborg, an AI-enhanced wingman program, which was soon followed by the Defense Advanced Research Projects Agency (DARPA) demonstration of the X-61A Gremlin program, which allowed for the deployment and recovery of swarms of small, sensor-laden UAS from cargo planes in flight" (Losey, 2022). On June 24, 2021, the USAF conducted its first live test of a program called Skyborg, which could one day "use AI system to operate uncrewed robotic wingmen that would accompany traditional fighter jets into battle" (Verger, 2021). According to a USAF spokesman, its goal is "for the artificial intelligence software to be able to modularly pop into different types of uncrewed aircraft and autonomously aviate, navigate, and communicate, and eventually integrate other advanced capabilities" (Verger, 2021). The additional capabilities would undoubtedly include the software necessary for UAS to fly alongside or ahead of manned fifth- and sixth-generation fighters, or MUM-T. The Skyborg UAS could fly ahead, conducting intelligence, surveillance, and reconnaissance (ISR) of the battlefield, conducting electronic warfare, or simply serving as a weapons carrier with missiles and bombs. Their most significant advantage would be that they are unmanned, cost far less than manned fighters, and are expendable.

Other nations are also working on pairing a manned fighter with an unmanned partner. The partner would be used for scouting ahead of the manned fighter to detect and receive attention from enemy air defense systems and possibly fighter aircraft utilizing stealth to hide from US pilots. The Royal Australian Air Force refers to their Loyal Wingman program as the Airpower Teaming System. In March 2021, the Australian Ministry of Defense awarded Boeing a $115 million contract for three additional Loyal Wingman UAS. Not all active and retired Air Force senior leaders are on board with the concept of an advanced, artificial intelligence-enabled Loyal Wingman. However, the supporters believe it is a logical and even inevitable evolution for air combat. USAF has been working on the next generation of highly stealthy fighter aircraft for years to replace the F-22 Raptor air superiority fighters.

In 2002, the NASA Dryden Flight Research Center conducted an autonomous aerial refueling test between two Navy F/A-18A fighter aircraft to demonstrate the models for the automated aerial refueling of UAS. In 2007, the Dryden Flight Research Center, Northrop Grumman, and DARPA successfully demonstrated autonomous aerial refueling. Two unmanned NASA Global Hawk aircraft were used to demonstrate unmanned to unmanned aircraft refueling. In 2015, the US Navy successfully transferred over 4,000 pounds of fuel from a manned aerial tanker to Northrop Grumman's X-47B UAS (Helfrich & Rogoway, 2023). In March 2023, an Airbus A310 Multi-Role Tanker Transport (MRTT) aircraft assumed control of an autonomous UAS in flight and refueled the aircraft. The first few tests were designed to demonstrate what many air forces are working toward, MUM-T. Airbus is building enhanced AI navigation sensors and algorithms to allow for autonomous UAS to fly dangerous missions in place of manned fighter aircraft and, when required, to team up with the manned fighters to conduct ISR, battlefield interdiction, and air superiority. These demonstrations are part of the European development project called the Future Combat Air System (FCAS), an endeavor similar to the USAF's NGAD program (Helfrich & Rogoway, 2023). These two programs are focused on developing advanced manned aircraft that will network with and control unmanned UAS aircraft.

DARPA has been deploying and recovering UAS in mid-air for years. Its experiments are helping USAF quest to build a fleet of autonomous, AI-enhanced UAS that would accompany manned fighters into combat. For the past two decades, the USAF has operated UAS, such as the MQ-9 Reaper, with impunity throughout large swathes of the globe to gather ISR. That impunity ended in 2019 when Iran used an air defense system to shoot down a $160 million RQ-4A Global Hawk surveillance UAS over the Strait of Hormuz (Victor &

Kirkpatrick, 2019). In 2023, Russia sent two SU-27 fighter aircraft to down a $28 million MQ-9 Reaper that was flying in international airspace over the Black Sea as it was undoubtedly conducting ISR of Ukraine and Russian forces (Liebermann et al., 2023). In what could be the next war against China, considered a near-peer adversary, the US should expect multiple areas where it operates freely today to be contested, including the high seas, airspace, the cyber realm, and space. The USAF UAS wingmen could perform the following tasks: provide situational awareness and designate targets for the manned aircraft controlling the UAS with infrared, electromagnetic, radar, or visual sensors; serve as communications nodes for friendly forces; conduct electronic warfare operations; fire air-to-air or air-to-ground munitions; lastly, it could serve as a decoy to force the enemy to deplete its air defense missiles while protecting the manned aircraft. The USAF plans to team the Wingmen, such as the XQ-58A Valkyrie or MQ-20 Avenger, with its sixth-generation or NGAD fighter. Still, they could also be teamed with the current operational fleet of F-22 and F-35 fighter aircraft and the new strategic stealth B-21 bombers. The B-21 has been engineered for both manned and unmanned missions and will control UAS accompanying it on missions. Former Secretary of the Navy Ray Maybus stated that the "F-35C will be the last manned fighter ever to exist" (Osborn, 2023a). This is due to the emergence of unmanned systems and their impact on survivability, forward surveillance, and even precision weapons attacks. The US military and CIA have used precision UAS attacks for decades, most notably in Iraq and Afghanistan. Yet they retain human control when decisions about lethal force against humans are required.

USAF has been studying the idea of low-cost, autonomous UAS, referred to by the Service as Loyal Wingmen, that could fly alongside its next-generation fighter aircraft for years. Now they are examining if such Loyal Wingmen could be controlled by air battle managers typically located in purpose-built aircraft referred to as Airborne Warning and Control System (AWACS) or possibly in the Air Force's large fleet of refueling tankers, the KC-46 Pegasus or E-7 Wedgetail (Losey, 2023). According to General Brown, USAF Chief of Staff, the Loyal Wingmen's missions could include "striking targets, conducting intelligence, surveillance and reconnaissance missions, or electronic warfare operations such as jamming enemy signals," as well as others (Losey, 2023). The benefits of Loyal Wingmen would be that they would be far less expensive than manned aircraft since they would not require a human crew. Depending on the mission and specific capabilities of the unmanned aerial system selected, they would be reusable but inexpensive. The variety of UAS would encompass a range of sizes, capabilities, and expendability.

Image courtesy of Trock, https://www.msn.com/en-us/news/technology/ skyborg-the-drone-that-could-revolutionize-air-force-operations/ar-AA1cL2bL?cvid=c99f5d9998a 44e20977fd3500bda8ab0&ocid=winp2 fptaskbarhover&ei=12

The AFRL has developed Skyborg, a UAS designed to accompany manned aircraft, provide surveillance, and absorb possible enemy air defense strikes if necessary. Skyborg will use its sophisticated cameras, sensors, and AI algorithms to provide a real-time view of the aerial battlefield as it unfolds. Skyborg's AI algorithm will be able to highlight to the manned aircraft "points of interest in its vicinity, enemy aircraft, missiles, unknown entities, obstacles, or allies in distress" (Trock, 2023). With the help of Skyborg's vigilant eye, pilots will have a view of everything around their planes and significant points of interest to avoid or act upon. The Wingmen accompanying the B-21, the F-22, the F-35, and the new sixth-generation fighter aircraft must be reusable but inexpensive. The USAF and DARPA are working on delivery methods, such as accompanying the manned fighters into harm's way, being stored inside the manned aircraft, and deploying when entering hostile airspace, or even in the holds of cargo or refueling aircraft that deploy them before contested airspace so they can converge with manned fighter aircraft. The Wingmen would be equipped with AI systems that could act autonomously if a threat emerged or if the manned control aircraft were downed, even making decisions such as shifting its preprogrammed targets.

The US Navy recently released a Request for Information, step one in the DOD procurement process, designed to solicit information from the defense sector to see if private companies could develop a system.

Image courtesy of Satam, https://www.eurasiantimes.com/us-hunts-for-hunter-drones-that-can-operate-from-non-carrier/?amp

The US Navy is looking for Aircraft Launch and Recovery Equipment (ALRE) to launch "fixed-wing, low-cost, reusable, and attritable, Group 3 UAS that would be launched from various naval vessels (Satam, 2023b). Group 3 UAS weigh between 55 and 1,318 pounds and fly under 287 miles per hour at an altitude of 18,000 feet (Satam, 2023b). These UAS would perform important roles such as ISR, logistics, and kamikaze attacks. The system for Group 4 or

5 systems could also serve alongside manned fighters or unmanned aircraft as platforms from which they would launch tens to hundreds of kamikaze UAS. Putting these UAS on multiple classes of naval vessels would allow naval commanders to use them as decoys to distract enemy attention, including their UAS and missile attacks. According to sources, the US Navy's Naval Sea Systems Command is close to awarding a $1 billion contract for kamikaze UAS (Hambling, 2023c). It is unknown if this is tied to the ALRE effort or is a separate initiative, but it has been designated, without competition, to be awarded to Raytheon. US Navy documentation states, "The Navy is pursuing a commercial-off-the-shelf solution with autonomous government-provided software and a government-provided launcher system" (Hambling, 2023c). Raytheon has developed existing systems to satisfy the Navy requirements and other loitering munitions called the Miniature Air Launched Decoy (MALD), which the Navy would use to confuse the Chinese in the event of war. This would allow the Navy to operate inside their extensive keep-out missile range based onshore installations.

US-based General Atomics recently unveiled its Eagle Eye multi-domain surveillance radar systems for the US Army, which can detect and track sUAS and will be installed on its Gray Eagle UAS. Eagle Eye's synthetic aperture radar can spot targets up to 50 miles away with high resolution and up to 125 miles at sea (McFadden, 2023b). The system's capability to identify sUAS is exciting to militaries worldwide as UAS has become widespread. UAS have demonstrated their wartime effectiveness against tanks, entrenched personnel, and fixed gun platforms in active combat zones like Ukraine and Israel. The threat from UAS will increase as technological advancements such as AI and machine learning improve their abilities and reduce costs. Gray Eagle UAS can locate the target, track its movements, and then share the target information with other platforms that can shoot the UAS with a cheaper weapon, such as a laser, kinetic weapons firing lead into the air, or a directed-energy weapon.

The US-based company Exosonic has undertaken an initiative to fill a hole in training opportunities for advanced, fifth-generation fighters to build supersonic commercial aircraft. The Revenant would be a supersonic UAS that could mimic enemy fifth-genera-tion fighters, the deadliest hostile fighter jets that US pilots are likely to encounter. The prospects were sound enough to convince senior leaders inside the Pentagon to fund this unique start-up company. The Revenant will possess a customizable, "highly modular plat-form with active electronically scanned array (AESA) radar, Infrared Search and Track (IRST) capability, and high bitrate data links [high bitrate refers to the higher amount of data transmission per second; more data is transmitted with a high bitrate and it requires more bandwidth and ensures better video quality] similar to what the F-22 Raptor carries" (Kirk, 2023). With a cruise missile design, stealth, and speed, the Revenant could conceivably be carried under the B-52s and air-launched, similar to how the current B-52-launches AGM-129 cruise missiles (Kirk, 2023).

Image courtesy of Kirk. https://www.autoevolution.com/news/new-american-supersonic-uav-can-mimic-russian-and-chinese-fighters-can-pack-aesa-radar-208046.html

Israeli manufacturer Steadicopter recently unveiled the Golden Eagle, a helicopter-type UAS that can be used for surveillance, recon-naissance, and striking enemy targets. The manufacturer claims that the Golden Eagle is the first unmanned helicopter capable of preci-sion strikes from the air.

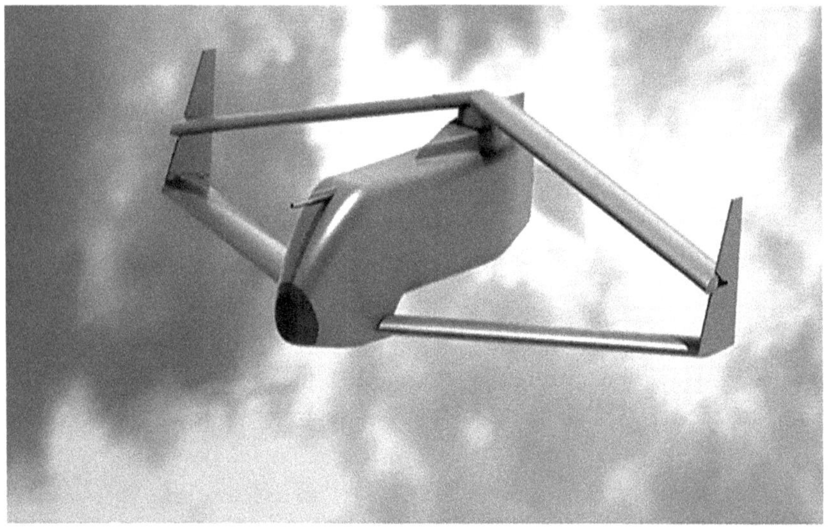

Image courtesy of Niedercorn. F.: https://www.armyrecognition.com/defense_news_march_2023_global_security_army_industry/sofins_2023_fly-r_launches_its_new_r2-120_raijin_loitering_munition.amp.html

It is equipped with advanced AI that simultaneously tracks and classifies multiple targets and comes with assault or sniper rifles. It has a maximum takeoff weight of 110 pounds, can fly up to 78 miles per hour, cruise at 53 miles per hour, and can use its electric motors to climb to an altitude of 18,000 feet (Panasovsky, 2023c).

A French company has developed a new generation of loitering UAS capable of carrying a three-pound explosive using a rhomboidal foldable wing design. The R2-120 Raijin is a tube-launched and is powered by an electric motor, allowing speeds of up to 120 miles per hour, 162 miles per hour when in terminal dive to its intended target, an operating range of 30 miles, remaining airborne for forty-five minutes, and can fly up to 9,000 feet in altitude (Niedercorn, 2023). The Rajinis is a unique, compact UAS; the aerodynamic drag is reduced, which enhances the aircraft's efficiency. The structural mass of the wing is decreased by a third, leading to a lighter aircraft; the rhomboidal wing supports a broad range of speed capabilities, making it versatile in different flight conditions; it maintains a high and relatively constant lift-to-drag ratio, improving its overall performance; lastly, the design of the rhomboidal wing allows for exceptional maneuverability, providing pilots with greater control and agility during flight (Niedercorn, 2023).

In a worldwide first, a British collaboration by BAE and Malloy Aeronautics launched a torpedo from a heavy-lift T-600 UAS, which boasts a payload of 440 pounds, a top speed of 90 miles per hour, and a range of 50 miles (Allison, 2023). The demonstration occurred during the recent NATO Robotic Experimentation and Prototyping with Maritime Uncrewed Systems (REPMUS) exercise to test the latest maritime technologies demonstration. The T-600 is a proto-type for the T-650 follow-on model that will be multimission capable, ranging from military missions such as casualty evacuation and anti-submarine operations to commercial involving automated logistics to humanitarian missions (Allison, 2023).

Image courtesy of Allison, https://www.ukdefencejournal.org.uk/british-drone-releases-torpedo-at-sea-for-first-time/

QinetiQ, a United Kingdom (UK)-based defense technology company, has been working on a highly secretive 3D-printed del-ta-wing kamikaze UAS for the Ministry of Defense. Defense offi-cials had established a competition of prototypes for testing equip-ment, systems, and technologies from five companies. QinetiQ has played an essential role in developing new UAS technologies for the UK Royal Navy, including producing the jet-powered Banshee Jet 80+ that has been launched from the deck of the aircraft carrier HMS *Prince of Wales* as a target (Newdick, 2023b). The Banshee has demonstrated the potential for it or its successors to serve as operational platforms aboard the UK's two aircraft carriers. The UK acknowledged that it transferred "at least 2,000 UAS to aid Ukraine, including loitering munitions, cargo-carrying logistics drones, autonomous mine-hunting vehicles, as well as 850 hand-launched Black Hornet micro-drones, the latter as part of a deal in collaboration with the Norwegian Ministry of Defense" (Newdick, 2023b).

Image courtesy of Newdick, https://www.thedrive.com/the-war-zone/clandestine-u-k-program-developed-3d-printed-suicide-drone-for-ukraine

A Chinese company is developing a UAS that will be launched underwater, exit the water, and fly as a kamikaze UAS, providing close air support to soldiers on the ground. It is called the TJ Flying Fish, and its weight is 3.6 pounds; it has a six-minute hover time and a speed of 6.5 feet per second while swimming underwater (Knutsson, 2023). It would be useful if China decided to invade Taiwan since the People's Liberation Army and Naval marine units would land with support via the UAS. The strategy would be for "hundreds of long loiter time UAS, they can provide a form of suppressive fire for an amphibious landing. The idea is, with hundreds of UAS flying over an area, the enemy troops cannot get out from cover or maneuver, allowing the attacking side freedom of movement" (Satam, 2023a; Knutsson, 2023). This type of requirements-driven solution provides a glimpse into how radically different amphibious landings might occur in the near future.

Image courtesy of Knutsson, https:www/foxnews.com/tech/creepy-chinese-drone-swims-underwater-flies-air

The US Navy is also developing a trans medium unmanned system that can operate underwater as a UUS but can also fly in the air as a UAS.

Image courtesy of Trevithick (2023c), https://www.thedrive.com/the-war-zone/drones-that-swim-and-fly-to-be-launched-recovered-by-uncrewed-submarine

The small UAS Naviator is scalable depending on its mission, has a 16-foot wingspan, and can transport payloads up to 90 pounds (Trevithick, 2023c). The Naviator will be able to conduct search and rescue missions, map the location of underwater mines, conduct surveillance against enemy ships, and inspect underwater bridges, dock piers, ship hulls, and oil drilling platforms.

Civil unrest

The US company that produces Tasers has plans to place them on unmanned aerial systems to help stop school shootings, but whether that occurs is up for debate as Axon is facing a backlash from its board of directors on the viability and ethics of creating such a system. Meanwhile, Axon says that the high-tech solution is necessary amid a debate on gun policy in the US. It has begun the development of a miniature Taser and the targeting algorithms required to allow operators to hit their intended targets (Debusmann, 2022). Axon plans to include a series of "integrated camera networks that allow schools and businesses to share real-time access to sensors with local public safety agencies" (Debusmann, 2022). While a novel solution, Axon cannot decide who would operate the Taser-equipped UAS. One solution would be for the systems to be equipped with encrypted keys that could only be accessed, thereby used "by police departments, federal agencies or even Axon employees" (Debusmann, 2022).

Federal Aviation Administration

As the global commercial UAS industry grows and matures, international aid organizations will inevitably acquire these delivery UAS. Commercial UAS, at least in the US, requires approval from a slow-moving FAA. It has reluctantly accepted the reality of a UAS environment and has moved painfully slowly to develop guidelines for its acceptance and operation within the US airspace. The legal determination that a UAS operating at 15 feet above the ground peering into a teenage daughter's room equates to the same legal guarantees and protections of an airplane is ridiculous. The father of the teenager who stops filming his daughter changing her clothes and knocks the UAS with a bat out of the sky could face federal charges for disrupting the safe operation of an aircraft. Depending on how the UAS is stopped, there are likely to be state charges and possibly a civil suit by the owner-operator of the UAS for damaging or destroying his UAS. The US has 1.3 million registered drones and over 116,000 registered operators. The legal consequences for disrupting, interfering, or destroying a UAS is as follows:

- "Destroying, damaging, or interfering with a drone is a federal crime, as drones are classified as aircraft and bound to the same laws.
- Under federal law, Title 18, US Code 32, the destruction of aircraft is a federal crime. The Federal Aviation Administration (FAA) classifies all types of aerial drones as "aircraft."
- Even recreational flyers are required to follow FAA regulations for flying in uncontrolled airspace.
- The Code aims to protect drones from any damage and destruction. The penalty for shooting at drones is illegal as per US federal code 32 and could result in a fine and 20-year imprisonment.
- Although you may own the property, you don't own the airspace above it. Throwing anything at the drone to knock it out of the sky is also illegal.
- Jamming any type of radio signal is illegal.
- Federal law prohibits using and selling jamming equipment that interferes with any authorized radio communications.
- This includes trying to jam the signal between a drone and its operator. Spoofing is another method used to disrupt everyday communication between the drone and the operator.
- Spoofing is illegal in the United States. Spoofing involves using software or hardware to spoof the signal used to communicate with the drone.
- This allows the spoofer to take control of the drone.
- Shining a laser pointer at a drone is illegal and may not interfere with its communication.
- Pointing a laser pointer at an aircraft is a violation of federal law and potentially state laws.

- As a drone is considered an aircraft, pointing a laser at it could result in jail time and fines" (FAA, n.d.; Aero Corner Editorial Team, n.d.).

The FAA has been dragging its feet for years in the quest by delivery companies to operate their UAS beyond the visual line of sight (BVLOS) of the operators. With trees and other obstructions in neighborhoods, it would be tough for a company to quickly deliver goods to multiple houses if they are hamstrung by having to keep the visual line of sight with their UAS. The industry will flourish by allowing companies that have demonstrated the hardware and software safety protocols to operate beyond the visual line of sight. When this occurs, the likelihood of battlefield replenishment of humanitarian supplies will become a reality. Recently, the FAA granted Florida-based power utility Percepto authorization for their autonomous maintenance UAS to fly BVLOS. Surely, commercial deliveries are next.

Air delivery

UAS in the US are classified into five categories based on their weight. Categories 1 and 2 are sUAS and weigh less than 55 pounds. Category 3 UAS are medium-sized and weigh between 56 and 1320 pounds. Categories 4 and 5 are large UAS weighing more than 1,320 pounds. Most "sUAS belong to Categories 1 and 2 and are mainly used for intelligence, surveillance, and reconnaissance missions or as attack platforms with a first-person view" (Edwards, 2023a). For commercial deliveries, which are beginning to occur at select test locations in the US and overseas, most UAS required to deliver goods will be Category 2 and 3.

Walmart recently announced expanding its UAS delivery service in Arizona, Arkansas, Florida, Texas, Utah, and Virginia (Guggina, 2022). Walmart estimated that UAS deliveries would reach an estimated four million customers and deliver one million packages annually within thirty minutes of placing the order (Guggina, 2022). There are initial restrictions: UAS deliveries will only occur between 8 AM and 8 PM; orders cannot exceed 10 pounds; the delivery fee will be $3.99 (Guggina, 2022). The stores that offer DroneUp deliveries will house a team of certified UAS operators who will conduct operations within FAA guidelines. Walmart has agreed with UAS companies Zipline and Flytrex to operate the UAS delivery operations for them. Walmart will also offer UAS options, including insurance, emergency response, photography, and real estate (Guggina, 2022). Walmart contends that it is merely keeping up with changing customer shopping habits, thanks to the worldwide pandemic and the added convenience that UAS deliveries provide.

Taxi service

San Jose, CA-based United's Midnight is building what it believes will be the answer for urban taxi service in the twenty-first century.

Image courtesy of Archer Aircraft and Rains. msn.com/en-us/travel/news/united-s-midnight-air-taxi-is-on-track-to-debut-by-2025-take-a-look-at-the-plane/ss-AA1c1Hb3?ocid=winp2fptaskbar&cvid=afd7d28dce184edbc3b0cbcaf160629d&ei=23#image=1

It is scheduled to fly its taxi's maiden flight by 2025 and has already received a $1 billion order from United Airlines. The electric vertical take-off and landing vehicle (eVTOL) is optimized for short-haul missions over traffic-congested cities. In March 2022, the FAA changed the certification category for eVTOL to "power-lift because although it flies forward like a regular airplane, it takes off and lands like a helicopter; this meant that the criteria for certification changed because it was no longer qualified as a regular airplane" (Rains, 2023). As such, few regulations were specific to power-lift aircraft, so the FAA had to create new regulations.

Image courtesy of Delta Airlines and Rains, https:///www.businessinsider.com/delta-makes-a-60-million-investment-in-evtols-joby-aviation-2022-10

Not to be outdone by United Airlines, Delta Air Lines invested $60 million in Joby Aviation's entry into the eVTOL taxi service, which is a five-seat air taxi that is expected to enter service in 2024, a year earlier than Archer Aviation's taxi (Rains, 2022). The purpose is to avoid traffic congestion by quickly flying between city centers and airports.

Chinese manufacturer EHang recently announced the preorder of its largest UAS taxi by Japan-based helicopter service provider AirX (Mircea, 2022a). AirX ordered 50 units of the dual-seat autonomous air taxi EH216. The EH216 is designed as a two-passenger taxi with a maximum payload of 485 pounds. With the maximum payload, it can fly for up to 21.7 miles, attain speeds of over 80 miles per hour, and fly to an altitude of over 9,800 feet. It uses 16 electric motors, has two gull-wing doors, and requires two hours to recharge its battery (Morcea, 2022a) fully.

Image courtesy of https://www.autoevolution.com/news/ehang-autonomous-aerial-vehicle-manufacturer-lands-its-largest-air-taxi-pre-order-in-japan-179736.html?utm_source=ae_self&utm_medium=ae_moreon &utm_campaign=ae_moreon_news_static

All the known extra-large UAS designed to operate as taxis possess VTOL capabilities, so they do not require runways or airports to work from. Instead, they can land or take off from small pads separated from the local populace throughout metropolitan areas, making them far more efficient and convenient for urban users. EHang is developing a longer-range electric two-seat UAS called the VT-30, which the company claims will have a range of up to 186 miles (300 kilometers) on a charge (Morcea, 2022). EHang has ambitions that the large Japanese preorder will spur additional "urban air mobility (UAM) projects in Japan to include air taxi services for the 2025 World Expo in Osaka" (Morcea, 2022).

In late 2021, Kawasaki completed a proof-of-concept testing of its VTOL K-Racer X1 UAS (Searles, 2022). The X1 can transport a total payload of 220 pounds, is unaffected by manmade road traffic or geographical features, and comes equipped with a robotic loading and unloading mechanism separate from the UAS (Searles, 2021).

Image courtesy of Searles, https://www.roboticsandinnovation.co.uk/news/deliveries/kawasaki-trials-cargo-vtol-and-robotic-ground-crew.html

The X1 would be ideal for areas where the geography is difficult for normal delivery vehicles to traverse, such as jungles, mountains, or remote islands. The proof of concept is intended for the logistics sector, where humans can be eliminated from many physical and mundane jobs. At the same time, robotic devices do the more dangerous and menial tasks associated with the loading and unloading of transport UAS. Once proven here, Kawasaki no doubt intends to develop more robotic devices to load and unload semi-tractor trailers and other delivery vehicles that are on the roadways all over the globe.

Cargo UAS

San Diego–based Natilus is focusing on upending the freight business in a big way. It claims its autonomous aircraft will reduce carbon emissions by up to 50 percent and costs by 60 percent while offering a bigger payload capacity (Mircea, 2022b; Ramirez, 2022).

Image courtesy of https://www.autoevolution.com/news/natilus-announces-6-billion-in-purchase-commitments-for-its-autonomous-cargo-aircraft-181354.html#agal_4

ZeroAvia is developing multiple engines, including a 2.5 MW hydrogen engine for the extra-large UAS, and is expected to be available by 2026 (Panasovskyi, 2023e). Since 2016, Natilus has been working on developing its fleet of fully autonomous cargo UAS that it says will revolutionize the aerial transportation industry. Natilus recently announced more than $6 billion in advance purchase commitments or 440 preorders for its UAS. Nautilus offers four types of remotely piloted cargo aircraft, each with its own capabilities: a 3.8-ton payload short-haul UAS, a 60-ton payload medium and long-range model, and a 100- and 130-ton payload long-range UAS (Ramirez, 2022). Some are suitable for domestic flights while others are engineered for intercontinental transport. All models feature a blended wing body design that offers 60 percent more cargo volume; the smallest of the four has a payload of 8 tons while the biggest boasts 130 tons of payload and is designed for long-range flights (Mircea, 2022b; Ramirez, 2022). With its next-generation autonomous aircraft, Natilus aims to offer sea freight and air transportation the best attributes. International cargo is currently moved via air, which is expensive, and by sea, which is thirteen times more affordable but fifty times slower in delivery. Nautilus will offer the speed of aerial delivery at the cost of sea delivery. The plane's exterior shape is significantly more aerodynamic; thus, it can fly faster while carrying more internal cargo. Natilus has completed two wind tunnel tests to validate its aircraft and expects to achieve the first flight of a full-scale prototype in 2023.

The US Bell Aircraft conducted live flights of its Autonomous Pod Transport (APT) 70 cargo UAS through the Dallas-Fort Worth airspace simulating a critical medical transport mission (Tegler, 2022). The FAA approved the test flight and was a component of "the National Aeronautics and Space Administration's (NASA) Systems Integration and Operationalization demonstration, which seeks to foster progress toward commercial use of unmanned aircraft systems" (Tegler, 2022). The demonstration also exercised the APT 70 systems responsible for navigating the US airspace among manned aircraft and the reliability of its command-and-control system. Bell is developing a family of cargo UAS for "package deliveries, disaster relief, offshore resupply, and other logistics missions for civil and military users" (Tegler, 2022).

Image courtesy of Tegler, https://www.forbes.com/sites/erictegler /2020/10/02/bell-learned-that-its-apt-70-drone-can-integrate-with-manned-traffic-during-a-demo-flight-what-did-it-learn-about-avoidance-and-5g/?sh=2cf5b3622elf

The APT 70 appears to be an old-style box kite that measures 6 × 9 feet with its rotors. After takeoff, the APT 70 tilts for forward flight. It has a cargo pod inside its four booms that can carry up to 70 pounds of payload. The APT 70 has a range of 35 miles and a top speed of 115 miles per hour.

Swarms

The threat posed by UAS transitioned from one or two to swarms quite some time ago. In a nutshell, UAS swarms can "coordinate their actions to accomplish shared objectives" (Zoldi, 2021). A swarm of UAS could be used to defeat enemy air defenses as part of a military operation or for humanitarian purposes such as search and rescue and disaster response. Obviously, the greater the number of sensor-equipped UAS in the air, the more data points or information there are for decision-makers.

Spectrabotics LLC, a US company specializing in software tools to aggregate, integrate, and analyze drone data," is focusing its analytics capabilities on using UAS swarm technology to help solve environmental issues such as hazardous materials (HAZMAT) spills. Spectrabotics uses high-definition video from UAS to provide an assessment of the environment; UAS equipped with thermal imaging cameras provides the location of HAZMAT hot spots; spectral sensors map the extent of the spill; and finally, light detection and ranging (LiDAR) provides intelligence on the extent that the HAZMAT spill could grow if left unattended. Spectrabotics has developed software that receives, sorts, integrates, and analyzes multiple inputs from UAS into a single coherent assessment.

The technology to allow a UAS swarm to coordinate individual movement has grown immensely in a short period. Software developers have been programming autonomous capabilities into UAS flight control systems, and the result of a growing number of UAS swarm demonstrations is impressive. This software that allows swarming capabilities is vulnerable to hacking and other cyber-related tools. An innocent swarm show celebrating the opening of the first Hyundai Genesis dealership in Shanghai, China, in April 2021 demonstrated the world's most extensive use of a UAS swarm to showcase the event (Hyundai Motor Group, 2021). The means to take over swarms or reprogram UAS in flight is a significant concern. The cyber tools necessary to conduct such nefarious operations are daily growing in complexity and capabilities. This new means of statecraft could potentially put the control of millions of UAS into the control of a terrorist or an adversary nation. In 2021, the DOD released the first strategy compiled by the executive agent for the DOD on this topic, the US Army, to widespread self-congratulatory fanfare. In the age of modern warfare, the use of UAS as an offensive system dates back to its use by Israel in the 1980s, so forty-plus years after its emergence is hardly an adequate response or timely achievement. What was especially significant was the impact that the DOD had on the passage of US legislation, specifically H.R. 5515 115th Congress John S. McCain National Defense Authorization Act for Fiscal Year 2019, that granted the DOD the domestic authority to use a range of defensive measures up to and including force, to defend specified personnel and installations from UAS (U.S. Congress, 2019).

The cybersecurity issue discussed here regarding UAS swarms only touches the surface of a more significant issue of cybersecurity vulnerabilities in the US. The infamous 2020 Solar Winds cyberattack infected thousands of organizations worldwide, including the US federal government, and led to widespread data breaches. These cyberattacks exploited "vulnerabilities in an IT monitoring and management platform resulted in undetected and unauthorized access to thousands of networks and systems through routine, trusted software updates" (Oladimeji and Kerner, 2022). The attack highlighted supply chain vulnerabilities, but with the global market being what it has evolved into, there are likely technologies or components whose origins are from a potential adversary, such as China. The People's Republic of China (PRC) National Intelligence Law's (as amended in 2018) chapter 2, "Authority of National Intelligence Work Institutions," article 7, states that "all organizations and citizens shall support, assist, and cooperate with national intelligence efforts in accordance with [the] law, and shall protect national intelligence work secrets they are aware of" (China Law Translate, 2017). In simple terms, the Chinese intelligence service can order any Chinese company to turn over data gathered on its products from any country. This is why governments worldwide have blocked the equipment from the Chinese telecommunications giant Huawei from providing hardware for next-generation mobile networks known as 5G. The fear is that purpose-designed backdoors could allow constant monitoring of all communications transitioning through the Huawei network equipment. A similar concern is directed toward DJI, the world's leading manufacturer of small UAS, with a global market share of 76 percent in 2021 (Schmidt & Vance, 2020).

Turkey

In early 2000, Turkey aimed to build its modern and self-reliant defense industry. The results have been impressive: the Bayraktar TB-2 UAS has gained fame from its use by Ukraine against the Russian forces but also in Libya, Syria, and the Azerbaijan-Armenia conflict, and as a result, it has been exported for use across the globe. As a result of the TB-2 notoriety and numerous international deals by Turkey's 2000 defense companies and other far less famous weapon systems, Turkey's global exports rose by 42 percent between 2020 and 2021. It is focused on raising its $4.4 billion in military exports to $6 billion in 2023 (Turak, 2023). The weapon sales have increased Turkish foreign influence around the globe, in Kazakhstan, Kyrgyzstan, Turkmenistan, Azerbaijan, Poland, Saudi Arabia, Qatar, United Arab Emirates, and Tunisia. Allen Ginsberg stated, sometime between 1965 and 1971, "War is good business. Invest your son." The unprovoked Russian invasion of Ukraine and Ukraine's defiance have created a vast global demand for arms. European countries that were chastised by then US president Trump for their repeated failures to spend 2 percent of their pledged Gross Domestic Product (GDP) on defense spending to bolster their dwindled military capabilities are at the forefront of those nations looking to purchase new weapon systems, especially air and missile defense, and long-range artillery systems as part of their lessons learned from Ukraine, as well as stockpile known ammunition and other military supply items that will be depleted quickly in the vent of hostilities. Turkey's defense companies have been inundated with military orders as the second largest military force in NATO. They are booked for years in advance to fill current orders.

In early October 2023, a Turkish company, Aselsan, conducted a test of a swarm of their Albatros-S USS against a ship in the

Mediterranean Sea (Ozberk, 2023). The test involved eight Albatros-S USS attacking a fifty-foot ship detected by a TB-2 Bayraktar UAS, which passed the target location to its headquarters. They, in turn, issued the attack command to the Albatros-S USS, which then swarmed the target, with one ramming the target ship resulting in the warhead in the nose to detonate and sink the target ship. The Albatros-S consists of a naval platform that is 4 feet wide and 23 feet long, it can attain speeds of 40 knots, weighs 4,310 pounds, and can carry a payload of 551 pounds (Ozberk, 2023).

Ozberk, T. (2023). SINKEX: Turkiye Demonstrates Swarm Of Kamikaze USV Capabilities. https://www.navalnews.com/naval-news/2023/10/sinkex-turkiye-demonstrates-swarm-of-kamikaze-usv-capabilities/

Turkey's first homegrown aircraft carrier, the Türkiye Cumhuriyeti Gemisi (TCG, "Republic of Turkey Ship") *Anadolu*, will be a crucial asset for its naval strength and will improve its ability to conduct missions abroad. *Anadolu* will also be the first carrier in the world to use UAS for its air wing. *Anadolu* is a 757-foot-long, 105-foot-wide ship that weighs 27,436 tons and can travel at 20.5 knots, cover 9,000 nautical miles, and stay at sea for fifty days (Brimelow, 2023). *Anadolu* was initially planned to carry F-35 fighter jets, but Turkey was kicked out of the F-35 program in 2019 after it purchased Russia's S-400 air and missile defense system, which raised security concerns among the US and NATO about the F-35.

Image courtesy of Brimelow, https://www.businessinsider.com/turkey-preparing-for-first-aircraft-carrier-tcg-anadolu
-with-drones-2023-2

Turkey then turned to its UAS industry to equip *Anadolu* with fixed-wing aircraft. The UAS that will operate from *Anadolu* is the Bayraktar TB3, a naval version of the successful Bayraktar TB2 made by Baykar, a Turkish defense company. The TB3 will have several improvements over the TB2, such as a higher takeoff weight, foldable wings, and a larger payload capacity. This will enable the TB3 to carry more weapons and fly farther than the TB2, with a similar endurance of about 24 hours. Baykar's Kızılelma, a jet-powered combat UAS still under development, may join *Anadolu*'s air wing. The Kizilelma fighter drone will have a range of 500 nautical miles, a ceiling of 35,000 feet, a payload of 1.5 tons, and a speed of 0.6 Mach and will also have advanced radar and stealth features, such as an angular shape and internal weapons bays (Brimelow, 2023).

An Estonian robotics firm recently acquired by a UAE defense company has successfully combined the UAE kamikaze UAS with the eastern European unmanned ground vehicle. The THeMIS Combat USS, developed by Milrem Robotics, was equipped with the Hunter 2-S loitering munition, made by Halcon., thereby allowing the manned squad to attack targets deeper in enemy territory. The Hunter 2-S is the first UAE-made UAS designed to swarm with other UAS and share data and communication. The UAS has a 4.4-pound payload, an 800-foot cruising altitude, a 56-mile-per-hour speed, and a forty-five-minute endurance (Helou, 2023).

Monitoring critical infrastructure (electrical power)

Today Percepto, creators of award-winning autonomous robotics technology, announced its upcoming deployment of hundreds of drone-in-a-box (DIB) systems to support the largest power company in the United States, Florida Power & Light (Zoldi, 2022). This arrangement constitutes the largest commercial autonomous UAS project in the world. The two pioneers intend to transform critical infrastructure resiliency for Florida Power & Light's 11 million customers through it. DIB systems involve autonomous UAS that launch from and return to networked landing boxes. These solutions allow organizations to prestage equipment onsite for scheduled or on-demand remote independent operations. They also automate workflows and provide large volumes of high-resolution data in unparalleled quantities.

During phase one of this project, which will occur later this year, Florida Power & Light will introduce 13 Percepto DIBs across the West Palm Beach area to monitor substations and power distribution grids. Over the next five to seven years, during follow-on phases, Florida Power & Light will similarly inject hundreds of Percepto UAS through a singular network across the entire Sunshine State. This latest effort builds on prior collaborations between Florida Power & Light and the Israel-based autonomous industrial inspection tech leader. Florida Power & Light launched Percepto's DIB technology for research and development in 2018. At the time, Florida Power & Light was the first to integrate autonomous UAS into state-wide utility site surveillance and inspections. Based on that initial project's successes and safety record, the FAA recently issued Florida Power & Light the first-of-its-kind nationwide beyond ground-breaking approval, enabling Florida Power & Light to fly Percepto UAS BVLOS at Florida Power & Light-owned or serviced sites. Current FAA rules require commercial drone operators to fly their aircraft within their visual line of sight. In November 2020, Percepto rolled out its autonomous inspection and monitoring DIB solution, the first-ever enterprise solution for on-site autonomous robot fleet management. With Percepto autonomous inspection and monitoring,

UAS can be operated remotely to ensure they work in sync, providing maximum coverage at the sites they monitor. Autonomous inspection and monitoring also automatically generate relevant business insights for stakeholders and decision-makers through advanced data collection and management features.

Disaster relief

Percepto recently announced an improved Autonomous Inspection and Monitoring system equipped with a new DIB called Air Mobile, a more compact, flexible, and lightweight UAS. Other improvements include AI and advanced machine learning, resulting in autonomous inspections and monitoring of primarily residential areas. As a result, Percepto claims that its operational costs and downtime have been shortened while increasing its efficiency and safety.

Florida Power & Light is initiating a new plan to alleviate the consequences of hurricane disasters by leveraging its technologies using DIB, other machines, and increased visual sensors to monitor even more of its assets. On average, hurricanes impact Florida twice yearly, resulting in hundreds of millions of dollars in damage and widespread power outages to up to five hundred thousand customers. As soon as the storms depart, DIB goes into action, surveilling large swathes of Florida to assess damage and severity, directly aiding the deployment of maintenance teams to bring power systems back online. The UAS are continuously flying and conducting surveillance, providing real-time status updates to the Operations Center, resulting in a far more efficient, prioritized distribution of repair assets and a return to normalcy for the power company and its millions of customers.

Nefarious use in the US

According to an unofficial estimate compiled by Judicial Watch, the Mexican drug cartels have launched more than 9,000 UAS across the border in the past year alone (Davis, C., 2022). The UAS are being sent daily to drop illegal drugs, scout routes for human trafficking, and to spy on US law enforcement and Border Patrol personnel. One of the more nefarious uses of UAS has been in the prison arena. There has been an increased use of UAS to smuggle drugs, weapons, and other contraband into prisons, not just in the US but worldwide. In just the last year, UAS operations smuggling contraband into prisons have caught the attention of law enforcement in multiple states (Daleo, 2023). In 2020, the US Federal Bureau of Prisons data showed 57 UAS incidents reported by prisons in the preceding year while the Department of Justice reported 170 UAS incidents between 2015 and 2019 (Daleo, 2023). These agencies have been slow to react to the procurement and installation of C-UAS systems at US prisons, plus the overall budgets of agencies and the financial support required to install C-UAS systems at hundreds of prison facilities do not exist today. One of the major problems with this, aside from a slow and bureaucratic procurement process by the federal departments and agencies involved, is the speed at which UAS advancement occurs. For each C-UAS system, some criminal elements and techies will outspend the federal government's efforts to get ahead and stay ahead of the threat. The US federal government only recently acknowledged the danger posed by UAS and has never and probably will never catch up to the threat. For comparison, according to information from the Correctional Service Canada, the Collins Bay prison had a mere ten UAS incidents involving the delivery of contraband five years ago, 247 in the last five years, and almost 100 in 2022 alone (Daleo, 2023).

Firefighting

A Nebraska company, Drone Amplified, received a $1 million grant in 2020 for "research and development from the National Science Foundation and Nebraska Department of Economic Development" and developed a novel approach using UAS and dragon eggs to initiate prescribed burns to aid firefighters in their yearly battles with forest fires (Osipova and Thorbecke, 2022). According to the company CEO and founder, Carrick Detweiler, one of two University of Nebraska at Lincoln engineering professors who left academia to form a startup company, the technique refers to the "controlled application of fire by a team of experts to reduce hazardous fuel in areas prone to wildfires; it creates "a very low-intensity burn that will burn up the dead leaves and sticks that would cause major wildfires when they dry out later in the summer" (Osipova and Thorbecke, 2022). The use of prescribed burns dates back centuries and was used by the Indigenous Americans and is still used by modern firefighters who travel by foot, in vehicles, or via helicopters.

Photograph courtesy of Osipova and Thorbecke, https://amp-cnn-com.cdn.am pproject.org/c/s/amp.cnn. com/cnn/2022/11/17/tech/drone-amplified/index.html

Detweiler states, "About a quarter of all wildland firefighting fatalities are related to aviation" (Osipova and Thorbecke, 2022). The cost efficiencies are noteworthy; for the same cost as a helicopter, about $80,000, firefighters can deploy hundreds, possibly thousands of the Drone Amplified systems. The company's UAS carries up to 400 dragon eggs, fireballs filled with potassium permanganate and glycol that ignite when the impact on the ground mixes the two chemicals. Using UAS armed with payloads of these novel dragon eggs will allow firefighters to work safely from the fires, reduce helicopter crews' exposure to operations at low altitudes over the raging fires, and increase the precision of initiating multiple precision burns nearly simultaneously. It can fly over rugged terrain with limited visibility without risking human lives and is controlled via an app on any firefighter's cell phone.

Counter-UAS

Former US Central Command chief Marine Gen. Kenneth McKenzie Jr. referred to cheap and commercially available UAS in the Middle East as "the most concerning tactical development since the rise of the improvised explosive device in Iraq. These systems are inexpensive, easy to modify, weaponize, and proliferate. They provide adversaries the operational ability to surveil and target US and partner facilities while affording plausible deniability and a disproportionate return on the investment, all in our adversaries' favor" (Keller, 2023). The Russian–Ukraine War has demonstrated that UAS are now integral to modern warfare and that current C-UAS systems are not widely available, effective, or cost-effective. In 2021, Israel sent two F-35 aircraft to kill two Iranian UAS approaching its airspace. The F-35 costs $100 million each, and the fiscal year 2020 hourly operating cost was $41,986 (GAO, 2022). A PATRIOT missile fired to kill an inbound UAS is approximately $4 million, and a defensive UAS designed to search for an inbound UAS can cost between $30,000 and $150,000 (DSCA, n.d.). Their targets would be readily available commercial UAS rigged to drop explosives that cost as little as $200 to as much as $10,000. Russian and Iranian-supplied UAS have damaged 40 percent of Ukraine's critical infrastructure, specifically its power stations. ISIS and Iran have used UAS to target US soldiers in the Middle East. The worldwide increase in the use of UAS as offensive systems is primarily due to their low cost, especially compared to manned aircraft, cruise, and ballistic missiles. This uniquely cheap and asymmetric threat has created headaches for developing an effective, cost-effective, layered defensive network. Directed energy systems are highly effective at countering small numbers of UAS or swarms at a low cost per engagement, require less energy than laser C-UAS systems, are not affected by bad weather or cloud cover, and are deadly against the next generation of counter-UAS systems such as UAS that do not require active links to its operators or possess anti-jamming coding.

A Greek company, Spirit Aeronautical Systems, successfully affixed and test-fired a 70mm Hydra 70 rocket from one of their quad UAS called the SARISA SRS-1X. The intent is to be a highly versatile and cheap replacement for manned helicopters capable of providing organic close air support for military units. The company plans to test-fire an FZ275 LGR laser-guided missile from its UAS in the coming months (Panasovskyi, 2023b).

Another Greek invention is a 300-kilowatt counter-UAS system called Minotaur. Greek scientists rose to the challenge purportedly by the Turkish UAS incursions over Greek airspace. Turkey has not denied the overflights, and "in September, Ankara's state-run news agency *Anadolu* published a report accompanied by pictures with the title "Drones capture images of armed vehicles sent by Greece to the Aegean islands with non-military status in violation of international law" (Kokkinidis, 2022). The Greeks have been keen on developing their own indigenous-produced UAS industry but have thus far trailed far behind Turkey. Greece recently unveiled the Archytas, a fixed-wing VTOL UAS (Kokkinidis, 2022). The VTOL capability allows great flexibility in takeoff and landings, whether on land or ships. The Greek Navy operates "small helicopter-type drones, the Alpha 900," off their frigates (Kokkinidis, 2022).

One of the many lessons that observers have taken away from the Russia-Ukraine War is that critical infrastructure, in the case of Ukraine, principally power plants, distribution centers, and communications hubs, are ideal targets for an attacker. As the term critical infrastructure implies, targeting and disrupting essential infrastructure products will detrimentally affect a nation's ability to sustain itself, whether it is the energy sector, chemical, manufacturing, defense industrial base, food, agriculture, etc. There is a certain degree of irony in that Putin claimed, among other things, that he feared a NATO expansion eastward by accepting Ukraine as a member of the alliance, which factored into his decision to invade it. The opposite has been the effect of his decision; NATO has recognized that Russia remains a threat and has increased its defense expenditures and signed contracts for the modernization of various forces within its military. In addition, Finland and Sweden applied for NATO membership, and Finland was admitted in April 2023. Most, if not all, NATO member nations desire to purchase increasingly capable IAMD systems to protect themselves from cruise, ballistic, and possibly hypersonic missile attacks. Many NATO countries have purchased UAS and more importantly, C-UAS systems to protect their key facilities and critical infrastructure. Paying close attention to the targeting of critical infrastructure in Ukraine by Russia, NATO recently established a specially formed task force to determine the best courses of action to preclude this from occurring to its members in the event of an outbreak of hostilities with Russia.

In 2022, BAE Systems successfully tested ground-to-air tests of 70mm rockets guided by Advanced Precision Kill Weapon System (APKWS) kits against UAS that weighed 25 to 50 pounds and flew up to 100 miles per hour. A BAE spokesman stated, "Militarized drones are becoming more prevalent in conflicts around the world, and we're giving our customers an efficient way to counter them without wasting expensive missiles" (Inside Unmanned Systems, 2022). The US and other nations have used 70mm rockets in their attack helicopters, which are combat-proven and highly reliable. The standard 700mm rockets are retro-fitted with APKWS guidance kits and proximity detonation fuses, resulting in a precision, low-cost supersonic C-UAS munition armed with a 10-pound warhead.

To partially address the new threat posed by UAS, MARSS Group, a UK leader in technology and AI, built and installed a new hybrid intelligent system that teams humans, machines, and AI to protect a critical infrastructure site in the Middle East against the threat posed by UAS. The 2019 cruise missile and UAS attack against oil processing facilities at Abqaiq and Khurais in eastern Saudi Arabia. MARSS Group believes that by combining all of the installation's sensors and effectors, X-ray emitters, jammers, lasers, kinetic weapons, etc. into "a single, intelligent user interface," their system maximizes the performance of individual instruments. It minimizes the decision chain deliberation, hence the target detection to neutralization (Hill, 2023).

The USAF AFRL successfully demonstrated the capabilities of its Tactical High-power Operational Responder (THOR), a high-powered microwave emitter designed to knock out inbound UAS and swarms of UAS in defense of air bases. THOR is self-contained inside a twenty-foot container and can be operational within three hours. THOR aims to increase the keep-out range while providing multiple engagement opportunities, unlike current C-UAS guns, nets, and laser systems.

The Air Force Research Laboratory's Tactical High-power Operational Responder (THOR) developed for airbase defense. (Courtesy photo/AFRL Directed Energy Directorate)© Provided by Task & Purpose

A lesson learned from the Russian invasion of Ukraine and Ukraine's efforts to C-UAS is to examine cheaper alternatives to expensive air defense engagements via ground-launched missiles to kill hostile UAS. One possible answer is through the development and use of microwave-emitting C-UAS systems. Epirus was recently awarded a $66 million contract by the US Army to build prototypes of Leonidas, a directed energy or electromagnetic pulse (EMP) system. Unlike kinetic and laser C-UAS systems, Leonidas can search for and engage an entire swath of airspace, and through its use of gallium nitride chips instead of vacuum tubes, power in radar and directed energy beams is amplified. Leonidas is remarkable in combining machine learning for frequency identification. It can adjust its energy output, frequency, energy pulses, and other aspects of the EMP depending on the threat UAS it has identified and targeted (Tucker, 2023).

Image courtesy of Epirus, www.defenseone.com/technology/2023/01/new-microwave-weapon-answer-iranian-drones/382111/

The US DOD is reportedly interested in a project called the Autonomous Multi-Domain Adaptive Swarms-of-Swarms (AMASS), a highly technological weapon of mass destruction (Study Finds, 2023). The goal of AMASS is "the ability to plan and execute missions that utilize thousands of autonomous entities in the degradation or defeat of adversary anti-access/area denial (A2/AD) capabilities," which is what China has been focused on for some time in their planning for an invasion of Taiwan (McMillan, 2023). This proposed capability could involve hundreds or thousands of UAS, USS, and UUS collaborating to engage enemy strongholds. The unmanned systems would carry kinetic kill devices, target-identifying GPS, and radar jammers. Human control of such a swarm or swarms of swarms would not be possible and would require an elaborate communications network, AI, and autonomous software to control. If the conflict in Ukraine is an example, AMASS would be used to attack critical infrastructure and military targets and even target human beings located in underground, concrete-reinforced command-and-control bunkers.

During the 2023 inauguration of newly elected president Luiz Inacio Lula da Silva in the capital of Brazil, four suspicious UAS approached (Makichuk, 2023). The Brazilian secret service was equipped with DroneGun Tactical on that faithful day. This C-UAS technology disrupts the electronic signals within a UAS, bringing them crashing to the ground without incident.

Image courtesy of Drone Shield, https://davemakichuk.substack.com/p/danger-of-deadly-drone-attacks-rears

The manufacturer of DroneGun Tactical, Austrian-based company Vornik, believes that the $10 billion market will continue to expand as the threat posed by UAS in public areas continues to proliferate, well beyond what was only a few years ago in the realm of terrorists, nonstate actors, and drug cartels. Vornik has sold its C-UAS systems to the US Air Force, French Army, Tokyo Police, and military and law enforcement agencies in Australia (Makichuk, 2023).

Anduril Industries recently unveiled an autonomous jet-powered UAS interceptor designed to destroy aerial threats ranging from large UAS to manned aircraft.

Image courtesy of Anduril Industries and Bogaisky, https://www.msn.com/en-us/travel/news/billionaire-palmer-luckey-unveils-a-new-jet-powered-drone-and-it-looks-bonkers/ar-AA1kOTbq?cvid=8b4323ef712d49868f158ca6ba105be4&ocid=winp2fptaskbar&ei=23&sc=shoreline

The Roadrunner is powered by twin turbojet engines that enable it to quickly get to its target. It remains in a vertical climate-controlled box that the company calls a Nest, which shields it from the environment or harsh weather conditions. Roadrunners are equipped with a warhead and will self-destruct when they intercept an inbound threat, or if the threat is not confirmed, they can return and land for reuse. Christian Brose, the chief strategy officer, stated "The requirements that we built into Roadrunner were focused on addressing the threat, where it was going and then where we believe it's going to continue to metastasize beyond where it is today" (Bogaisky, 2023). Multiple Roadrunners can be launched and managed by a single operator and are capable of autonomously determining their flight paths and intercept courses. The company is planning to produce nonkamikaze Roadrunners with different types of payloads, such as electronic warfare equipment and even to fight forest fires by dropping fire suppressants.

Chapter 4

Unmanned Surface Systems

Military applications (USS on land)

Since 1995, the US has used UAS to perform surveillance missions; later, it was armed. The Russia-Ukraine War has accelerated the development and use of unmanned systems, UAS, USS, and UUS. It is estimated that 11,000 UAS are being used globally for high "altitude reconnaissance, battlefield surveillance, and engagements with onboard munitions" (Shayotovich, 2023). Less advertised but equally important are small, handheld personal reconnaissance systems, small USS and UUS, and robotic mules that currently carry equipment for soldiers. US-based General Dynamics Land Systems has completed a version of a contender for the Army's Robotic Combat Vehicle (RCV) competition. The US Army has proposed three RCV variants from industry: a "Light version weighing less than 10 tons, a Medium sized one between 10 and 20 tons, and a Heavy robot weighing between 20 and 30 tons" (Roque, 2023).

Image courtesy of General Dynamics Land Systems. https://breaking defense.com /2023/03/gdls-showcases-short-range-air-defense-payload-on-tracked-robot-10-ton/

It is built with lightweight materials and a hybrid propulsion system allowing multiple platforms for different AI-enhanced Manned-Unmanned Teaming (MUM-T) missions. General Dynamics has focused its efforts on the base of the Tracked Robot Demonstrator, 10-ton (TRX) while collaborating with other companies in their areas of expertise. For example, the base of the five-ton TRX could be outfitted with various five-ton upper components specializing in a wide variety of battlefield requirements such as direct and indirect fire, autonomous resupply, complex obstacle breaching, short-range air defense (SHORAD) C-UAS, electronic warfare (EW), and reconnaissance missions.

Photograph courtesy of Oshkosh Corp. https://defence-blog.com/oshkosh-defense-announces-robotic-combat-vehicle-submission/

The photograph showcases the SHORAD variant equipped with eight Stinger missiles, a 30mm cannon, a 7.62mm machine gun, and assorted target tracking systems. This variant is already in use with the US Army onboard their manned Stryker-based SHORAD system. There are many lessons to be learned from the ongoing Russia-Ukraine War, one of which is that the age of unmanned systems is clearly upon us. If the US Army wishes to get ahead of its strategic competitors, it would be wise to speed significantly up its future force design and various competitions to incorporate AI and unmanned systems. US-based Oshkosh Corp. has also submitted its proposal in response to the US Army's RCV program based on leveraging existing technologies.

The BAE entry to the Army's USS competition is the Black Knight, which uses "color video cameras, Laser Detecting and Ranging (LADAR), FLIR thermal imaging cameras, and GPS, and they can plan a route across a battlefield while avoiding any obstacles it encounters" (Shayotovich, 2023).

Image courtesy of Shayotovich at https://www.msn.com/en-us/news/technology/this-futuristic-unmanned-tank-could-be-a-game-changer-for-the-us-army/ar-AA1gvk1I?ocid=hpmsn&cvid=444 04595cb5143 afbef8d151bff7 6233&ei=15

The Black Knight is one of the newest competitors to the US Army's program for a USS. It is powered by a 300-horsepower Caterpillar diesel engine with a top road speed of 48 miles per hour.

It allows the vehicle to be remotely controlled while they navigate the vehicle remotely. It is also equipped with a 25mm cannon and coaxial 7.62mm machine gun.

The US Defense Department has approved a trial of a USS made by South Korea's Hanwha Aerospace, named the Arion-SMET. It will undergo a three-week test at a US Marine Corps (USMC) training site in preparation for its deployment with the USMC.

Image courtesy of Hyeon-hwan, J. and Tae-gyu, K., https://www.msn.com /en-us/news/world/korean-defense-firm-to-test-unmanned-vehicle-for-u-s-marines/ar-AA1gIJSI?cvid=2b82898842c14d70b4dada0707b75236& ocid=winp2fptask bar&ei=19

Arion-SMET is a 2-ton vehicle designed to assist infantry units in various tasks, such as carrying weapons, transporting the injured, and providing close combat support. It has self-drive off-road navigation and a return-to-home function in case of communication loss. It can also be remotely controlled or operated autonomously for reconnaissance or surveillance missions. The US government has invested over $5 billion since 1980 to acquire foreign-made products through this testing program (Hyeon-hwan and Tae-gyu, 2023).

Boston Dynamics created the first known four-legged robot two decades ago and now, another company, Throwflame, has manufactured a robotic "dog that spews flames from its mouth," exemplifying the continued rise of automated capabilities (Sicard, 2023). In 2020, a UAS owner modified his system to shoot flames, which he used against a large hornet nest high up in a tree.

Image courtesy of Sicard, S. (2023). https://www. armytimes.com/off-duty/
military-culture/2023/07/14/this-flame-throwing-robot-dog-proves-were-all-
doomed/

The reason for the innovation with USS and USS is simple: in the US, there are no federal laws or regulations that would prevent a citizen from purchasing and using flamethrowers for reasons determined by the individuals as important. While not currently supporting military operations, it is easy to see that it will be in short order.

The USMC recently tested a USS that looks much like a mechanical dog with a "training version of the M72 infantry anti-armor rocket launcher" (Trevithick, 2023b). This armed walking USS is similar to one that the Chinese built using COTS, which costs between $2,700 and $3,500, depending on capabilities (Trevithick, 2023b).

Image courtesy of Trevithick, https://www.thedrive.com/the-war-zone/marines-test-fire-robot-dog-armed-with-rocket-launcher

The value of walking USS is immense: they could be armed with a variety of weapons to serve as scouts for friendly squads of troops. If necessary, it could engage the enemy target while notifying its higher headquarters of the location of enemy units. In addition, for urban areas and highly confined areas, such as tunnels, the use of walking USS would prove highly beneficial.

Cars and trucking industry

Gatik is a US-based company specializing in robotic technologies for autonomous trucks transporting various goods from distribution centers to stores. It has produced two autonomous trucks being tested on a closed, 7.1-mile route "between an e-commerce distribution facility there to a Walmart Neighborhood Market store," all without a human being behind the vehicle's wheel (Ohnsman, 2022). Autonomous driving technologies for local deliveries and the trucking industry are moving faster toward the services to market than on-demand "USS taxi services and personal vehicles" (Ohnsman, 2022). A great many companies are investing in and developing autonomous systems. TuSimple is focused on USS semis for long-haul highway trucking; Nuro is on USS that can deliver groceries and food delivery in neighborhoods; while Waymo is developing USS tractors and USS taxis (Ohnsman, 2022). A senior vice president at Walmart states, "The goal of driverless delivery in Bentonville is to fill e-commerce orders faster and increase asset utilization. Our work with Gatik has identified that autonomous box trucks offer an efficient, safe, and sustainable solution for transporting goods on repeatable routes between our stores" (Ohnsman, 2022).

Roads

Hybrid and electric automobiles have begun to change how people view going green for our roadways, but the biggest hurdle to broader acceptance has been the limited number and location of recharging stations. An Israeli startup, Electreon, has unveiled a possible solution to wirelessly charge moving vehicles via a retrofit program for the estimated 40 million miles of roads globally (CIA, 2021). To demonstrate the practicality of its product, Electreon fitted its wireless charging system under the asphalt of a closed loop track and drove a Toyota sedan for 1,207 miles at 30 miles per hour (Levin, 2023). In this test, only a quarter of the track had been fitted with wireless charging, but the company stated that even at highway speeds, the charging system can keep an electric vehicle fully charged. Electreon was initially focused on delivering wireless charging for buses and trucks that typically utilize limited and well-established routes but are looking to expand to the masses.

Image courtesy of Electreon and Levin, hhps://www.cnn.com/2023/03/14/politics/us-drone-russian-jet-black-sea/index.html

The trucking industry has autonomous semis on the interstates. Still, into and out of cities, a human would take control of them and guide them in congested areas and in and out of factories or other facilities.

Farms

In 1837, John Deere released the first commercial steel plow, helping to revolutionize the farming industry. In 2022, John Deere released the new 8R autonomous tractor, revolutionizing the farming industry again. The new 8R tractor utilizes six pairs of stereo cameras and AI to view the environment, navigate, plow the soil, and disperse seeds, all while avoiding obstacles, all controlled via a smartphone app by the farmer (Knight, 2022).

Image provided by John Deere and Knight, https://www.wired.com/story/john-deere-self-driving-tractor-stirs-debate-ai-farming/

Labor costs would be cut dramatically by autonomous self-driving tractors, which have been slowly inching into farm equipment. The "8R tractor relies on neural network algorithms to make sense of the information it collects, including data about the soil that will be used to improve its performance and to provide farmers with insights on how to work their land best (Knight, 2022).

The Ekobot WEAI USS is a four-wheeled, 1,322-pound, battery-powered weed picker that can operate 10–12 hours a day on a single charge, hit 2.5 miles per hour, and can clear 24.7 acres (Edwards, 2023b). The Ekobot uses GPS, safety sensors, and onboard

AI-enhanced visual systems to keep it focused on its tasks, which include using steel fingers to pull weeds. According to Ekobot, "its weed-plucking robot allowed farmers to grow onions with 70 percent fewer herbicides" (Edwards, 2023b).

https://arstechnica.com/information-technology/2023/11/mother-plucker-steel-fingers-guided-by-ai-pluck-weeds-rapidly-and-autonomously/

Military applications (USS atop the water)

The US Navy wants to take a page from the Iranian speed boats strategy to swarm US naval vessels with a twist: build small autonomous USS watercraft armed with Stinger antiaircraft missiles to protect its logistics ships initially and future expeditionary ships. The initial experimental USS will be based on a Greenough Advanced Rescue Craft (GARC) design, which is 15 feet and 8 inches long, 5 feet and 8 inches at its widest, possesses a low profile, has a 1,000-pound payload capacity, has a top speed of at least 35 knots, and a range of 400 to 700 nautical miles depending on its speed (Trevithick, 2023a). The Navy has been testing examples of the GARC for the past five years as a communications relay node and escort platform; the planned USS could be outfitted with various packages to include the multirole Mark 50 Gun Weapon System and systems capable of firing missiles. The GARC could also carry sensors, intelligence, sonar, anti-submarine, and sea mine packages. It would be logical for the GARC USS to be networked with others, thereby creating a distributed target detection, cueing, and engagement capability far greater than that of a single USS. The GARC could also be fitted with a "Towed Airborne Lift of Naval Systems (TALONS), which is a tethered parasail capable of providing an elevated platform for various equipment" or to serve as a communications relay (Trevithick, 2023a). If the Navy deploys a swarm of GARC, they could each be armed with different weapons, communication, and sensor systems depending on the threat platforms it would likely encounter.

There are many individuals sounding alarms on autonomous, AI-enhanced technologies. They fear a 21st-century arms race with AI at the center of it. While it is not an irrational fear, it is misplaced. A race was initiated years ago, and multiple nations are engaged in a winner-take-all competition. The first nation to achieve a "Hal-9000" degree of complexity, multitasking, and autonomy will be the world's dominant player in AI. Over a year ago, the US Navy demonstrated an autonomous engagement as it continues to move toward its goal of an armada of USS ships that uses AI to man, navigate, and fight adversaries without putting sailors in harm's way (Verma, 2022). China's threat to its manned ships drives the USS ship program. Specifically, China continues to build an arsenal of ship-killing missiles that it would use to keep the US Navy far out to sea when it decides to invade Taiwan, something that the US has said it would defend against. Human lives would be spared by putting a fleet of USSs in harm's way instead of manned ships.

There has been a great deal of criticism from industry experts, retired naval officers, and Congress on the Navy's ambitious plans for autonomous ships. The US, including its military Services, lacks a coherent and synchronized plan to develop AI and grow autonomous platforms. The Chinese do not. According to Dr. Kai-Fu Lee, a former Apple, Microsoft, and Google executive turned investor, China has a track record of getting things done and currently has every ministry from the Ministry of Science and Technology to the Ministry of Education focused on achieving its goal of AI dominance by 2030 (Lee, 2023). Even the US Government Accountability Office (GAO) stated that the US Navy underestimates the cost of building, outfitting, and deploying autonomous ships. For example, "if the Navy envisions autonomous vessels being out at sea for 30 to 60 days on end, AI software will have to do many things sailors normally would do to maintain the ship: keeping the ship's hull intact and performing oil changes. It will require a huge investment in this digital infrastructure, in the AI…to achieve that, and that's where we've seen underfunding on the part of the Navy" (Verma, 2022). In December 2020, the Navy released its thirty-year shipbuilding plan, and it contained provisions for "143 to be autonomous ships by 2045, with the first 21 to be ready by 2025 at an estimated cost of $4.3 billion" (Verma, 2022).

Recently, Thales conducted sea trials for its newest naval mine counter systems on behalf of the French and British navies. The new system included multiple sub-systems including integrated USS, UUS, sonars, AI, and cybersecurity to detect and neutralize sea mines. This is a "paradigm shift in mine countermeasures operations, ensuring naval personnel remains outside the mine danger area while advanced systems, backed by AI algorithms, enable real-time and post-mission sonar data analysis" (McNeil, 2023).

The USMC is developing a USS, what they refer to as a Long-Range Unmanned Surface Vessel (LRUSV), fitted with a UAS launcher with eight Israeli-manufactured Hero-120 loitering munitions (Trevithick, 2023). The LRUSV could launch the loitering UAS, extending its surveillance range and engaging any targets.

Image provided by the USMC. https://www.thedrive.com/the-war-zone/our-best-look-yet-at-the-marines-new-loitering-munition-toting-drone-boat

The Hero-120 would be used for ISR and could track and destroy targets on sea or land. The LRUSV would be semi-autonomous and could operate as a single entity or as part of a swarm. The Hero-120 can be carried in a single launch tube by the USMC on the ground or via a launcher. Their LAV-25 also carries wheeled light armored vehicles and is being tested on naval amphibious ships in groups of six. The Hero-120 has a maximum range of 37 miles, a 10-pound warhead, and can remain airborne for an hour. In 2022, the USMC demonstrated that a Hero 400 could be launched off a helicopter and controlled off a tablet by ground forces.

Bahrain and Israel are joining forces with other regional allies amid high tensions with Iran (Williams, 2022). This effort is designed to demonstrate an alliance against Iran and its nonstate proxies, including Yemen's Houthis rebels, attacking oil targets in Saudi Arabia and the United Arab Emirates to foment instability, chaos, and terror. The US Navy is weighing "adding unmanned Israeli boats to its joint Middle East operations, a move that could deepen Israel's military involvement in the Gulf" (Williams, 2022). The US Navy Fifth Fleet is based in Bahrain, and it is examining the various uses for unmanned surface, aerial, and undersea systems as part of its current naval exercise in the Arabian Sea. Across the entire gamut of systems, the Israelis are investing heavily in the whole spectrum of unmanned systems technologies.

Moving cargo

Hyundai accomplished a USS milestone with the competition of the world's first autonomous navigation across the ocean by the 134,000-ton commercial tanker, the Prism Courage (Cormack, 2022). It left the Gulf of Mexico and arrived in the Republic of Korea (ROK) thirty-three days later. The Prism Courage is a 980-foot-long liquefied natural gas carrier that sailed for approximately 12,427 miles thanks to an AI system installed by SK Shipping, a leader in autonomous navigation. The AI navigation system analyzes weather, wave heights, and nearby ships to determine the best course and adjust it in real time. The AI system encountered and avoided 100 other vessels in its voyage across the Pacific Ocean, increasing fuel efficiency by 7 percent and reducing greenhouse gas emissions by 5 percent (Cormack, 2022). Across the globe, autonomous navigation technology is a growth industry for both commercial and the public. For companies, the AI navigation system eliminates any workforce shortages by human beings; it has demonstrated that fuel efficiency was increased while emissions were reduced. Safety was improved by eliminating human error by removing dependence on human operators.

Suzaka

In May 2022, a Japanese cargo vessel weighing in at 749 gross tons displacement, named the Suzaka, became one of the very first autonomous commercial cargo vessels to complete 99 percent of a forty-hour 500-mile journey in the congested waters of Tokyo Bay without human control (Doll, 2022). It traveled from the port of Tokyo to the port of Tsumatsusaka and avoided 400-500 vessels along the way. Its artificial intelligence system was designed by Orca AI, an Israeli company, and the Suzaka provided real-time detection of other ships and natural obstacles, tracking, classification, and range estimates based on input from 18 onboard cameras, twenty-four hours a day (Doll, 2022). The real-time data from the autonomous ship was monitored by the company's fleet operations center in Tokyo, hundreds of miles away.

Zhu Hai Yun

In May 2022, China launched an AI-equipped USS that they say is designed for marine research that could easily be employed for military purposes. It is intended to serve as a hub for UAS, USS, and UUS, all of which could conduct surveillance or attack missions. The ship, the Zhu Hai Yun, is semi-autonomous and is 290 feet long and 46 feet wide, with a draft of 20 feet, a top speed of 18 knots, and a displacement of 2,000 tons (Trevithick & Parken, 2022). The manufacturer, Yunzhou Tech, demonstrated in 2018 that it could launch a 56-boat swarm to overwhelm a hostile ship in a highly coordinated manner. The Zhu Hai Yun could coordinate such swarm attacks.

Image courtesy of Trevithick, J. & Parken, O. (2022). https://sofrep.com/news/china-launches-ai-operated-mothership-zhu-hai-yun-for-drones-submersibles-and-boats/

Yara Birkeland

In 2018, Norway launched the world's first autonomous ferry. The *Yara Birkeland* is the world's first zero-emission autonomous cargo ship. The movement of the *Yara Birkeland* will eventually be monitored by three control stations located on land. It can transport 103 cargo containers with a top speed of 13 knots and is powered by a 7 MWh (7 megawatt hours equals 7,000 kilowatts of electricity generated per hour) battery (Beighton, 2021). This is the rough equivalent of one thousand times the capacity of an average electric car. Being crewless will save the company money on human crews. Initially, the loading and unloading will require human crews, but the company plans to increase the viability of the autonomous technology so that loading, unloading, berthing, and unberthing will be performed without human crews. Additionally, autonomous cranes and straddle carriers, vehicles that place the containers onto ships, will also be autonomous. At some point soon, these types of autonomous ships will have to start interacting with one another to exchange information and avoid conflict paths.

Image courtesy of Beighton, https://www.cnn.com/2021/08/25/world/yara-birkeland-norway-crewless-container-ship-spc-intl/index.html

The future for small to mid-sized USSs will undoubtedly be to resupply human populations on isolated islands and move people, vehicles, and supplies like ferries operate today.

Chapter 5

Unmanned Undersea Systems

Naval vessels

Vice Admiral (Ret) David Lewis stated, "Facing a tsunami of new technology and unsure how to adopt it, the Navy needs the fleet, an entity full of innovators and operations leaders, to find the opportunities and identify and help fix the problems. It needs new knowledge, created by sailors and Marines, to solve the newest suite of mysteries and accelerate the service forward. The Navy should be doing what works—bringing the operational fleets, Marines, and sailors into the technology innovation process. As it did in 1923, 1953, and many times before" (Lewis, 2023).

The US Navy submarine community is moving to build its UUS capabilities despite the surface fleet having gained the most attention for its research, development, and evolving use of unmanned systems. The submarine force has been funding two projects that appear to be close to competition: the Orca extra-large UUS test vehicle; and modifications that will allow the Razorback UUS to be deployed and recovered from a submarine's torpedo tube (Eckstein, 2022). "The XLUUV [Extra Large Unmanned Undersea Vehicles] is critical because it makes up, in some cases, for the lack of submarines.… It gives you additional capacity because you have a limited number of [attack submarines]," Vice Admiral William Houston, the commander of Naval Submarine Forces, told reporters at the Naval Submarine League's annual conference in November." (Eckstein, 2022). Once modifications and software enhancements are complete, the Razorback will allow "every nuclear-powered general-purpose attack submarine will have the ability to serve as a UUS mothership, Rear Admiral Casey Moton, the program executive officer for unmanned and small combatants" (Eckstein, 2022). According to "Rear Admiral Doug Perry, the Navy's director of undersea warfare on the chief of naval operations' staff, the attack submarine fleet has 200 torpedo tubes, and this development ensures every single one can be used to launch and recover medium-sized UUS if needed" (Eckstein, 2022).

The submarine community understands that possessing a UUS capability would give every attack submarine in the fleet the following abilities: to have a scout that would move ahead of the submarine and if lost, would not entail a loss of life nor would it give away the location of the submarine; second, the ability to execute third-party targeting; third, to aid in communications and in developing a real-time common operating picture for the commanding officers (Eckstein, 2022).

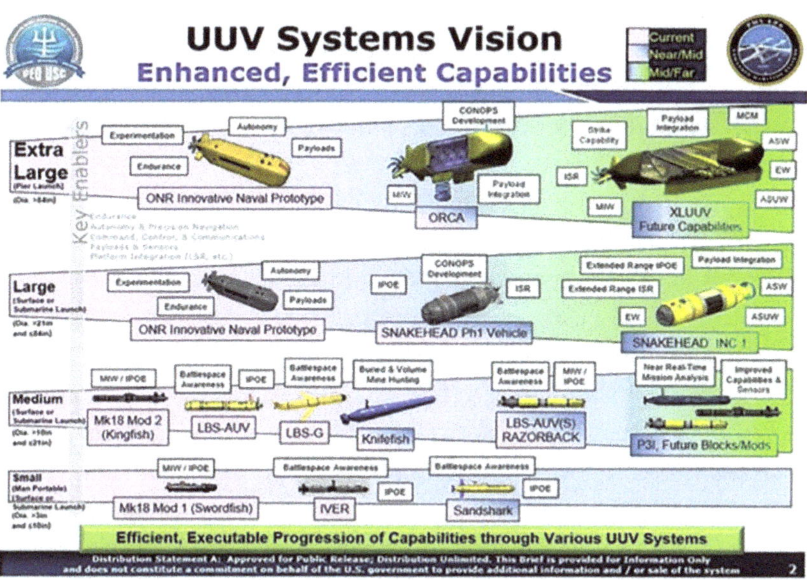

Eckstein, M. (2022). What's ahead for Navy unmanned underwater vehicle programs? https://www. defensenews.com/naval/2022/11/29/whats-ahead-for-navy-unmanned-underwater-vehicle-programs/

The Navy is also developing other UUS programs for small mine warfare and mine countermeasures. These programs include the following: "the Lionfish, which will replace the Swordfish UUS that has been in use for more than 20 years; the next generation Barracuda mine neutralizer, which will be untethered unlike the first generation; and the Medusa UUS, a mined, expendable, unmanned submarine asset as a replacement for the submarine-launched mobile mine" (Eckstein 2022).

Boeing won a $43 million contract for the fabrication, test, and delivery of four Orca XLUUVs and associated support elements. According to the US Department of Defense contract award notice, the Orca XLUUV will be an open architecture, reconfigurable UUS and will be modular in construction. The core vehicle provides guidance and control, navigation, autonomy, situational awareness, core communications, power distribution, energy and power, propulsion and maneuvering, and mission sensors. The Orca program will focus initially on payload integration work and mine warfare. Its future capabilities may include mine countermeasures, anti-submarine warfare, electronic warfare, antisurface warfare, ISR, and strike capabilities (Vavasseur, 2022)

In another sign of warming relations between the Arab world and Israel, Bahrain is openly cooperating with Israel primarily to counter increased tensions in the Arabian, also called the Persian Gulf, due to Iranian naval belligerence on the high seas (Williams, 2022). The maritime cooperation will lead to a normalization of bilateral relations between Bahrain and Israel. The bilateral plan is called the Warm Peace Strategy and prioritizes efforts for food security, water, sustainable energy, trade, investment, and later a joint tax treaty (Williams, 2022). The goal of the bilateral agreement is to build consensus amongst moderate Gulf region countries to provide stability, economic prosperity, and security against Iran and its support to the Yemen Houthi rebel faction, whose ongoing attacks against Saudi Arabia recently widened in the region to include the United Arab Emirate in 2022. Israel recently became the 29th country to post an attaché to the US Navy's Fifth Fleet headquarters in Manama, Bahrain, a move that could lead to the addition of Israeli unmanned naval vessels to the US Navy's ongoing Middle East operations. The Fifth Fleet is exploring the incorporation of USS into its Gulf exercises and is interested in Israeli systems to complement its growing inventory of UAS and UUS.

UAS, USS, or UUS can be critical in resupplying manned vessels at sea and in search and rescue operations. The future is bright, and the opportunities are abundant for those willing to fund this vital area of research and development.

Undersea

In 2022, the French Ministry of Defense released its strategy for dominating the ocean floors up to 6,000 meters below the surface with autonomous and remotely operated UUS. According to the Minister of Defense, Florence Parly, "the goal of the new strategy is to equip the French military with the ability to reach depths of 6,000 meters or nearly 20,000 feet. This makes it possible to cover 97 percent of the seabed and effectively protect our interests, including cables [laid on the seabed]" (Machi, 2022). The French Ministry of Defense and the British Navy are exploring the next generation of anti-mine warfare systems (SLAM-F), which will replace human divers with USS and UUS. Minister Parly also stated:

> That the seabed strategy as a natural progression. The three relatively nascent "domains" are all ripe for competition, thanks to advances in unmanned technologies. Today, the emergence of drones and remotely operated robots—driven by industry needs [and] capable of carrying out operations that meet military objectives at a depth of several thousand meters—are transforming the seabed into a new space for strategic competition. The sheer immensity and opacity of the deep-sea domain will make it difficult to characterize and attribute any actions that take place down below. But Paris must be able to react

to any such actions and be capable of protecting the critical seabed infrastructure like the fiber-optic cables that carry 99 percent of the world's digital data, she emphasized. Mastering the seabed will also help protect military and industrial secrets, Parly said. She cited the recent crashes of a US F-35 joint strike fighter in the South China Sea, and a British F-35 in the Mediterranean Sea, as proof "that the ability to search and recover a sensitive object from very great depths is not only a technical issue, but also a strategic one. France has the largest exclusive economic zone in the world, owing to its many territories still spread across the world—from islands like Guadeloupe and Martinique in the Atlantic, to French Guiana on the coast of South America, to Tahiti and the rest of French Polynesia in the Pacific, to mainland France itself in Europe. The seabed strategy is consistent with France's 2030 investment plan. The Ministry of Defense will be supported by the nation's recently reformed Ministry of the Sea, which has developed a strategy for the exploration and exploitation of mineral resources on the deep seabed. Government officials have earmarked 650 million euros (US $734 million) out of a COVID-19 economic recovery plan into the maritime sector." (Machi, 2022)

China has noted the importance of deep-sea mining for rare minerals necessary for our modern technologically based society and has five deep-sea mining contracts, the most of any country.

Chapter 6

Unmanned Outer Space Systems

An unmanned outer space system (UOSS) is uncrewed and explores outer space for scientific exploration. UOSS may be remotely controlled, guided, or autonomous from human control with a preprogrammed list of executable operations unless human controllers on Earth instructed otherwise. UOSS are typically fitted with scientific instruments and tools used to study the atmosphere and composition of space and other planets, moons, or celestial bodies and send their information back to Earth for analysis. A UOSS may operate far out in space, orbit, or land on a planet or a moon, make a one-way journey and transmit data back to Earth via radio, or bring samples and data back to Earth. UOSS must be able to withstand extreme environments to collect data. Voyager 1 spacecraft has now traveled farther than any spacecraft in history from our planet. It was launched in 1977 to conduct a fly-by of the planets Jupiter and Saturn. Yet it continued in its mission to gather and transmit data back to Earth and left our galaxy in August 2012. It is transitioning through interstellar space between galaxies into the great unknown. Sputnik 1 was the first manmade UOSS to go into outer space. It was launched on Oct. 4, 1957, by the former Soviet Union, and on Jan. 31, 1958, the US sent Explorer 1 into outer space. Both countries sent UOSS toward other planets. Mariner 2 was the first probe to study another planet, and on December 14, 1962, it flew past Venus. A different probe called Mariner 4 was the first probe to snap a picture of a planet. On July 14, 1965, Mariner 4 flew past Mars. Its images of Mars showed a cold, cratered, moon-like surface. In 1971, Mariner 9 arrived at Mars and became the first probe to orbit, or circle, another planet. Mariner 9 took a picture of Mars that showed the largest volcano in the solar system.

Once a UOSS has left Earth, its trajectory will likely orbit around the Sun, and its most practical method is a Hohmann transfer orbit. The maneuver was named after Walter Hohmann, the German scientist who published a description of it in his 1925 book *Die Erreichbarkeit der Himmelskörper* ("*The Attainability of Celestial Bodies*"), and transfers a spacecraft between two orbits of different altitudes around a central body. More complex techniques, such as gravitational slingshots, can be more fuel-efficient, though they may require the probe to spend more time in transit. A method using very little propulsion, but requiring a considerable amount of time, is to follow a trajectory on the Interplanetary Transport Network, which is a collection of gravitationally determined pathways through the Solar System that require very little energy for an object to follow by using Lagrange points, "points of equilibrium for small-mass objects under the gravitational influence of two massive orbiting bodies," as locations where trajectories through space can be redirected using little or no energy (Cornish, 1998). Due to lower cost and risk factors, many space missions are more suited to telerobotic rather than manned missions. Some planetary destinations, for example, Venus or Jupiter, are too hostile for humans to survive. Outer planets such as Saturn, Uranus, and Neptune are too distant to reach with current manned spaceflight technology, so telerobotics is currently the only way for humans to explore space beyond our moon.

Autonomous exploration until warp drive is invented

Deep space is a hostile and unforgiving environment with extremely harsh conditions for humans and hazards everywhere. The farther out a spacecraft goes, the less likely it will depend upon Earth's instructions to execute its mission and act successfully should the spacecraft encounter any unexpected hazards, such as an uncharted meteor shower. To build resiliency into future spacecraft, space agencies seek to exploit modern technologies to infuse adaptability and autonomy into their multimillion-dollar spacecraft. One such concept involves releasing a swarm of small, cheap craft designed to float on the solar winds toward the deep recesses of the Milky Way Galaxy. Swarming small spacecraft could cooperate with one another without specific centralized guidance from Earth operators and cover more significant volumes of open space far more efficiently than a single and costly manned spacecraft ever could. These tiny craft would be inherently cheap, robust, redundant, adaptable, and distributable for regions of space vs. one specific pinpointed location. When machine learning is added, small craft would break down tasks into their simplest form and then execute a strategy for efficiently accomplishing them. When swarm robotics are added, these small craft are interconnected and can share the processing of collected information and tasks and take actions autonomously without human input. The concept of a swarm of outer space systems adds flexibility by allowing the individual systems to collaborate and switch specific tasks. In addition, swarms provide for scalability between differing systems and numbers and provide for robustness, a likely situation to be encountered in the desolate reaches of outer space where meteor showers and other changes to the environmental conditions are possible or in the event of the inevitable loss of individual spacecraft.

Space exploration will begin with the moon as a lunar outpost and staging area for increased exploration into deep space, establishing human settlements. The moon will allow humanity to build the necessary infrastructure for a permanent human presence and mine and accumulate natural resources and materials. The experience and knowledge gained by a lunar outpost on our moon will serve as a foundational blueprint required for the eventual colonization of Mars and the movement into deep space. To date, over one hundred orbiters have been sent to the moon and forty to Mars (Kaczmarek, n.d.a). Rovers have been the go-to system to land and explore the uneven landscapes of the moon and Mars. For the first time in the history of space exploration, the Ingenuity UAS was launched from the *Perseverance* rover on Mars. It has flown over 51 missions, surviewing the landscape beyond the visual range of its rover and recharging its batteries via its onboard solar cells (JPL, 2023). The software and hardware advancements in rovers sent to Mars have been increasingly complex and astounding. The self-driving autonomous navigation system equipped with *Perseverance* has been truly impressive. "In 1997, the Sojourner rover needed to stop every 5.1 inches for its computer to reset before it could move again; the Spirit and Opportunity in 2004 could drive up to 1.6 feet before they needed to reassess their environment and move further; in 2012, Curiosity advancements eventually led to the two computer brain *Perseverance* which has set multiple records for its travels including autonomous travel without human control" (JPL, 2023). There is work being done on Earth for walking rovers, which would be able to explore lava tubes and other areas that current wheeled rovers cannot transit.

The age of space exploration is set to explode with countless opportunities. The space economy is "the full range of activities and the use of resources that benefit human beings by exploring, researching, managing, and utilizing space" (Kaczmarek, n.d.b). The space economy is increasing as the prospect of space mining, collecting rare earth materials from asteroids, long been a science fiction adventure, will soon it will be a reality. In 2020 alone, "the commercial space industry launched over 1,000 spacecraft into orbit" (Kaczmarek, n.d.b). Private industry is utilizing every efficiency to launch spacecraft cheaply and is posed to exploit the commercial opportunities in space, not government agencies. The space sector is increasing but also fueling growth in other sectors, such as telecommunications, satellites, energy, and transportation.

Both the moon and Mars missions have benefited from previous missions and the knowledge gained from them. Future missions to previously unexplored space and celestial bodies will require innovative AI and autonomous systems to handle the uncertainties and lack of understanding of the associated environments and dangers. These missions will also be operating autonomously for an extended period, depending on the speed of communications and locations of the nearest human outpost. Thus, these AI-enhanced autonomous spacecraft must be highly adaptive and learn rapidly. Work is being done on a small, autonomous, cheap spacecraft called SmallSat that weighs less than 400 pounds and can be launched into sectors of open space (Nesnas, Fesq, and Volpe, 2021). They will be capable of operating while transitioning to open space, navigating to celestial bodies and land, and continuing to send back critical information. The future of space exploration will be led by UOSS performing multiple tasks autonomously as directed by AI algorithms and doing so without human guidance.

A new study published in *Nature Synthesis* found that sending a robot equipped with AI could quickly conduct the necessary calculations to create oxygen for survival by humans on Mars, a task that humans would take a lifetime to complete such a task. According to the authors of the study:

"There are more than a million potential oxygen evolution reaction catalysts on Mars, which would give humans too many possibilities to work with when trying to create oxygen. "Oxygen supply must be the top priority for any human activity on Mars because rocket propellants and life support systems consume substantial amounts of oxygen, which cannot be replenished from the Martian atmosphere. Within six weeks, an AI chemist built a predictive model by learning from nearly 30,000 theoretical datasets and 243 experimental datasets. This was a demonstration of a robotic AI chemist for automated synthesis and intelligent optimizing catalysts for the oxygen evolution reaction from Martian meteorites. The entire process, including Martian ore pretreatment, catalyst synthesis, characterization, testing, and, most importantly, the search for the optimal catalyst formula, is performed without human intervention" (Lee, 2023).

AI-assisted chemistry has been applied to chemistry issues and has continued to make substantive advances. In 2020, AI helped researchers improve the production of hydrogen from water. This technology could help pave the way for human exploration into the universe. The authors further state "Our study demonstrates that an advanced AI chemist can, without human intervention, synthesize [oxygen evolution reaction] OER catalysts on Mars from local ores. The established protocol and system, which are generic and adaptive, are expected to advance automated material discovery and synthesis of chemicals for the occupation and exploration of the extraterrestrial planet" (Lee, 2023).

Chapter 7

Unmanned Systems in the Russia-Ukraine War

The Russian invasion of Ukraine and the Ukrainian armed resistance, thanks primarily to Western weapons and supplies, combined with innovation and superior leadership from the tactical through strategic levels, is an omen of how technology has shaped this war and will do so in future warfare. Ukraine has been creative in its use of military technologies. Its society is more accessible, flexible, and open to grassroots initiatives than Russia, and the Ukrainians have developed many homegrown military systems. Ukraine has also exploited technologies developed elsewhere, such as the Starlink internet service, US and British-manufactured UAS, artillery, missiles, and commercially available image-gathering equipment, resulting in a significant battlefield advantage.

The use of UAS has reshaped how expensive armored vehicles are viewed by those invested in them. The British Next Generation Light Anti-Tank Weapon (NLAW), made by Swedish firm Saab, costs $33,000 and has proven itself adept at destroying $10 million Russian tanks. The international supply of various UAS has proven to be as deadly as the quietest sniper, only instead of a long-range bullet being fired at a single target, the Ukrainians have adapted UAS for dropping small explosives on Russian soldiers, vehicles, communications towers, etc., and in doing so, they have instilled fear and desperation within the frontline Russian soldiers that are poorly trained and lead. Ukraine has increased its military might by developing its own

UAS and using them as an alternative to long-range jet fighters and missiles. It has begun launching long-range UAS to Russian targets over 100 miles inside Russia and is reportedly developing a model that will fly 600 miles. Their innovative use in adapting a commercial watercraft jet ski into a fast canoe-shaped high-explosive torpedo has damaged multiple naval vessels and threatened the remainder of the Russian Black Sea fleet. Their fear of these new devices has pushed the naval forces further out to sea and made them ineffective, both as a delivery means for amphibious fighting forces and direct support of those forces. Lastly, the era of AI on the modern battlefield has undoubtedly arrived and has been applied ingeniously by the Ukrainians. Kill chains with human echelons in the loop are notoriously slow. The enemy will often depart before the order to engage has come down from higher headquarters. The Ukrainians have shortened their kill chains by posting software engineers on the frontlines to support their fighting forces directly. By doing so, they can quickly calibrate algorithms to increase the effectiveness and accuracy of engagements against Russian forces.

The rise in the use and importance of UAS in the Russia-Ukraine War has been meteoric. It has shaken the foundations of strategies and resulted in a self-examination of militaries across the globe. This is the result of global commercialism and government recognizing the importance and various uses of UAS to hobbyists, businesses, and on the battlefield. If one merely looks to YouTube for videos of Ukraine UAS dropping explosives on Russian vehicles and soldiers, we will find them by the hundreds. There are estimates that Ukraine has "500 UAS in the air for every military operation it conducts" (Bowden, 2022). They were initially used for ISR of the battlefield and for spotting for field artillery engagements, something they continue to excel in today.

The value of UAS on the modern battlefield has been solidified by their daily use in the ongoing war between Ukraine and Russia, but large numbers of UAS are required to saturate the battlefield and have eyes on behind the forward line of troops. According to the Royal United Services Institute, "On average, Ukrainian quadcopters survive three missions, fixed wing UAS for six" (Bronk, Reynolds, and Watling, 2022). In Ukraine, competition has risen to see who can build a highly capable UAS for the least amount of money, often using readily available COTS products and 3D printers to create reinforced frames and wing structures.

Ukraine has utilized Soviet-era weapons and ammunition for use against Russia with surprising success. Years ago, the Soviet Union had armed its Ukrainian partner with Soviet Kh-35 subsonic anti-ship cruise missiles, which modern-day Ukraine has since modified and improved into the R-360 Neptune cruise missiles with increased range, targeting, and electronics (Trevithick & Rogoway, 2022). On April 13, 2022, Ukraine used them and a TB-2 Turkish UAS to successfully sink the Russian flagship Moskava while it was moored in the Black Sea (Trevithick & Rogoway, 2022). Ukraine recently reported that they are adding length to their Neptune cruise missiles for additional fuel, thereby increasing their range. Using Tu-141 and Tu-143 jet-powered reconnaissance UAS, Ukraine modified into cruise missiles and launched them against Russian airfields. All airfields with manned aircraft require large logistical bases and personnel to support them. Thus, they are large enticing targets for adversaries. Modern-day cruise missiles have superior ranges compared to manned aircraft and are considerably cheaper.

The worldwide unmanned systems industry, especially in Ukraine, has increased at a feverish pace. Early in the war, the Turkish Bayraktar TB-2 UAS were hailed as saviors of the Ukrainian combat troops. Able to approach the Russian forward deployed troops and even the flagship of the Russian Black Sea fleet, the Moskava, TB-2 rained havoc upon the Russians until their electronic warfare and jamming capabilities improved. Russia traded military assets and, no doubt, technologies with Iran and received thousands of UAS in return, which they quickly used to target Ukrainian critical infrastructure. The Ukrainians received hundreds of air defense systems and missiles and improved their skills and proficiency against attacking UAS, cruise missiles, and even hypersonic missiles. We have seen the deployment and successful use of loitering munitions by both sides, far more successfully by Ukraine, as well as its wartime development and use of UUS to target the Russian naval vessels in the Black Sea. According to Robert Clark, director of the defense and security unit at UK think-tank Civitas, "The industrialization of the drone industry in Ukraine is at a pace we have never seen before in peace or war. This is creating such an enormous proliferation of drone technology that it's completely changing elements of the battlefield" (Monks, 2023). Ukraine has initiated a campaign held outside Kyiv called an Army of Drones to allow companies to demonstrate their latest advancements while increasing the supply of UAS to the military. The campaign is part of an online crowdfunding initiative called UNITED24, has received hundreds of millions of dollars, and has dramatically expanded Ukraine's arsenal of UAS. UAS manufacturers use the campaign to demonstrate their wares in "ground target practice, chases, and air-to-air combat" (McLaughlin, 2023). Ukraine acknowledges that this aspect of modern warfare is highly technological and that this war they find themselves immersed in is very much technological. Ukraine has innovated and developed tactics, techniques, and procedures far faster than Russia. In addition, Ukraine has developed long-range UAS which it has used by attacking targets in and around Moscow, thereby bringing the war to the unsuspecting citizens of Russia's capital.

Technological advancements in unmanned systems, the focus of a mobilization of a nation's best and brightest, and the resolve to use them have allowed Ukraine to turn the tide of the war from within after depending on Western supplies initially. According to Dr. James Patton Rogers, from Cornell Tech Policy Institute, "There are small and medium powers that have not been able to acquire offensive air power capabilities who can do so now. This is what I call the second drone age. The West dominated the first, but now 113 nations have military drone capabilities…and at least 65 non-state actors" (Monks, 2023). This proliferation of unmanned systems has only increased the global threat potential. Immediately after 9/11, governments focused their internal counter-terrorism efforts on stopping terrorists from entering their countries. Unmanned systems, especially those enhanced with AI, could completely upend governmental efforts to counterterrorism. Unmanned systems give terrorists the ability of a terrorist to attack from a distance, another city, state, or country. Humans can no longer provide real-time oversight of immense data flowing into command-and-control centers. As AI is new and there is great hesitation in the West to allow it to operate without direct human management, those in Russia and China do not. What began as a retaliation effort to counter the vast numbers of Russian military platforms and soldiers, has the potential to alter our world more than anyone could have imagined.

Out of the Russian-Ukraine War comes another first by the Ukrainians; a heavy lift bomber UAS that drops what are improvised aerial bombs or carefully places 24-pound Soviet-era TM-62 anti-tank mines on the ground, effectively mining areas in front of, adjacent to, or directly behind Russian fighting positions (Hambling, 2023b). Ukraine has been adept at taking COTS UAS and converting them into bombers capable of dropping 30mm and 40mm grenades. As part of the Army of Drones initiative led by its digital transformation minister, Ukraine recently "procured 270 Vampire UAS capable of carrying up to 33 pounds over 4 miles, or 22 pounds to 6 miles (Hambling, 2023b). The Vampire has thermal imaging to allow the operators to guide bombs during night attacks, and the E620 Kazhan has also been optimized for night attacks with a payload of 44 pounds for short-range attacks (Hambling, 2023b). It is usually an 82mm mortar that serves as the bomb for these heavy-lift UAS. Mortar works well except when dropped on bunkers and concrete emplacements. For these targets, Ukraine is planning to unleash more powerful bombs, such as the TM-62 anti-tank mines, 80 percent of which is the explosive (Hambling, 2023b). Thousands of these Soviet-era mines are available to the Ukrainian military, whereas the Vampire and Kazhan UAS cost between $15,000-$30,000 each (Hambling, 2023b).

Autonomy is another lesson learned from the Russia-Ukraine War. Human operators must relinquish command and control to maximize the capability of kamikaze UAS and the sheer numbers that could be in the air at one time. The EW threat will be high. Thus, the UAS must not rely on links to their controllers and must have their electronics insulated against jamming; second, this would eliminate the need for many human UAS operators. Ukraine has successfully managed to deny its airspace to Russian manned aircraft, which failed to achieve air superiority early on and has never reinforced its initial failed efforts. Ukraine is regularly destroying ballistic, cruise, hypersonic missiles and most UAS that encroach into its airspace, thereby shifting the balance of power in the air to the defender.

The paradigm shift in warfare is best articulated by Dr. Hammes in one of his recent articles:

"Historically, transitions to new forms of warfare have taken decades unless accelerated by war. The transition from the large foot-mobile infantry divisions of World War I to the fully mechanized motorized divisions needed for World War II took two decades. The sift from battleship-dominated navies to carrier-dominated navies took just as long. In both cases, the outbreak of World War II accelerated the development of the concepts to exploit these new systems and the willingness to do so. Unfortunately, today's rate of change does not allow the United States and its allies and partners the luxury of two decades of transition to twenty-first-century warfighting. Nor do these nations possess the resources to upgrade all of their forces across the board simultaneously. Yet failure to do so places the US and allied forces at risk in any near-peer conflict. Historically, successful transitions require building game-changing capabilities while adapting existing forces to maximize their support to the new concepts" (Hammes, 2023).

Commercialism

Across the globe, commercial companies are focusing considerable effort and resources on figuring out how to create capable UAS delivery platforms. For the delivery systems to be effective, they must perform all the following: "vertical takeoff and landing, GPS navigation, electronic insulation since they will be transiting near cell towers and possibly airfields where the radio frequency (RF) energy bleed off could disrupt their internal electronics, reliability, low cost, endurance, and payload lift. For surveillance UAS, these same companies must incorporate multispectral imagery, long endurance, and payload lift dependent on the equipment required for the missions" (Hammes, 2023).

SYPAQ Systems, an Australian company, built and delivered to Ukraine their unique waxed cardboard Corvo Precision Payload Delivery System (PPDS) UAS, which can be easily assembled and launched with a catapult (McFadden, 2023a). These lightweight UAS have a flight range of up to 75 miles and have been designed to conduct surveillance, drop small explosives, or deliver ammunition, food, and medicine to soldiers on the front line. The Australian government launched an initiative and an AUS $1.1 million government contract to develop and supply drones. The PPDS costs a few thousand per unit, is expendable, and is shipped flat on a pallet with some unfolding required before use. Thus far, the Ukrainian military has used the PPDS more than sixty times as they are designed to operate autonomously in challenging conditions and are guided to their target via GPS, which is comprised of a constellation of satellites in synchronized orbit above the Earth that provide real-time position information. The control software will continue to fly the UAS to its intended target even if the GPS signal is lost or jammed.

The ongoing war between Ukraine and Russia is not only being carefully studied by nations across the globe, but it is serving as a testbed for new technologies and systems. Each nation is uniquely addressing the lessons learned. China has not fought a land battle since it was soundly defeated by Vietnam in 1979. Taiwan is paying attention and is reexamining its strategy for defeating a mainland Chinese invasion. It has invested heavily in 3D printing, and can 3D print multiple defensive parts and systems, including automatic rifles should China successfully blockade inbound shipments in wartime (Boyes, 2022). Iran has become Russia's primary supplier of UAS and is developing weapon systems and strategies for continued attacks against Saudi Arabia's oilfields and refineries, as well as for its proxies to conduct warfare against Israel.

In 1982, Israel became the first nation to successfully weld UAS to attack Syrian air defense units ahead of its manned fleet of aircraft. It has made significant advancements in UAS and USS production. In chapter 2, I presented specific information on the

Israeli micro drone that was designed to enter tight areas, conduct surveillance missions, and if ordered to do so, conduct strikes using a lethal charge equivalent to a grenade blast. Clearly, Israel is keeping a relatively low profile as it continues to develop and use UAs, even microsystems, against terrorists. Turkey gained acclaim in the years leading up to the Russia-Ukraine War, thanks primarily to its TB-2 Bayraktar UAS. In 2020, these UAS helped Azerbaijan defeat Armenia. It also gained worldwide attention when it began to supply Ukraine with them early in the war. Many people do not realize that the Ukraine company, Motor Sich, a turbine manufacturer in Zaporizhzhya, produces the engines for the TB-2. Russia has focused its Iranian-produced UAS kamikaze UAS on attacking this factory, though there are no reports of the factory being damaged or even closed. The ongoing media attention and purported successes of the TB-2 have fueled interest and sales worldwide. War has been good for UAS business. An estimated forty countries have UAS in their inventories, which are far cheaper than manned aircraft and easier to train operators on. Any teenager has already mastered the universal controller used in nearly all video games. The UAS are inexpensive to build and operate, and if they are shot down, another cheap replacement can be launched quickly. The use of UAS as surveillance platforms, loitering munitions, artillery spotters, and kamikaze tools have terrorized the young, poorly led, poorly equipped, and poorly trained Russian soldiers and they have provided pinpoint accuracy for Ukrainian attacks using artillery and armored vehicles. In addition, it is worth noting that before the war, Ukrainian military observers attended NATO exercises and took notes on being unpredictable and flexible in what Westerners refer to as the proverbial "thinking outside the box" approach that Western forces pride themselves on. Russian observers also attended NATO exercises but stated that their World War II tactics would remain unchanged. A war of overwhelming and direct force by armored columns rolling nonstop toward and over the defensive fortifications would remain. Unfortunately for the Russians, the Ukrainians mastered superior leadership, maneuver, and massing forces at decisive points across their armed forces. We have witnessed a small amount of USS action in the Russia-Ukraine

War, but their use and possible success by the Ukrainians primarily, has resulted in the Russian Black Sea fleet being moved far offshore. While they can launch surface-to-surface missile attacks, they are predominantly hitting, maybe even targeting, civilian dwellings and not Ukrainian military targets. Thus, they have remained high-value targets to the Ukrainians primarily for media purposes but have been relegated as insignificant and very expensive military platforms that must be kept at a safe distance from the Russians. They could hardly handle a second well-publicized sinking, such as the one involving their flagship, the Moskva, in 2022 by land-based cruise missiles that Russia provided to Ukraine many years prior, and Ukraine improved upon.

Another lesson from the Ukraine-Russian War is the speed at which an armed force engaged in combat can exhaust its ammunition and stocks of kinetic projectiles and fighting vehicles. Since the short-lived 2014 War against Ukraine, Russia has been testing cyber tools against Ukrainian military and commercial targets. At the outbreak of the second Russia-Ukraine War, Russia used cyber, massed military forces, and mercenaries to squash the Ukrainian will to fight. The senior advisors to Putin told him that it was likely that the Ukrainians would roll over and welcome the Russian army without resistance, much as they had done in 2014. This time the advisors were very wrong. The former comedian and current president, Volodymyr Zelensky, who had not been taken seriously by most global powers before the war, has emerged as a symbol of Ukrainian resistance. He has galvanized his people into fighting to the death while enticing Western countries to donate ammunition and weapons to the fight. If there were such a thing as a world presidential election, it is likely that Zelensky would be elected due to his demonstrated leadership, charisma, and ability to wield power and influence far more effectively than any other leader on the world stage. Early in the war, President Zelensky called for citizens to utilize their hobbyist UAS for surveillance and attack against the invading Russian forces. The responsiveness of his people surprised the Russians and possibly even Zelensky. He has mesmerized and embarrassed Putin at every opportunity.

Putin announced years ago that his country had developed hypersonic missiles, which were unstoppable. Zelensky convinced Biden to donate two batteries of PATRIOT air and missile defense systems. They quickly demonstrated that they could, and did, intercept six of Putin's highly prized Kinzhal hypersonic missiles aimed at them. They have also shot down Russian aircraft operating within Russian airspace. Russia's use of missiles and Iranian UAS to target areas where its air force would be shot down has reinvigorated the need for modern air defenses, both short and longer range. The US Army cut its SHORAD force at the start of the 1990 Iraqi War and reappropriated those personnel authorization slots to grow the Special Forces and Military Police units. Soon after the Iraqi War ended, the Army again reallocated its force structure thanks to their having to relearn the lessons of past wars. It began the long and tedious march to regrow SHORAD capabilities. If Putin aimed to show the West that NATO was getting far too close to Russian borders and that he needed to demonstrate the consequences of crowding Russia, then his gamble has failed miserably. What has transpired is that NATO has gained additional member nations, and all member nations have focused on their military shortfalls and pledged to fix them with dedicated funding and training. Many have purchased highly capable air and missile defense systems to protect their political and military targets and critical infrastructure; in Ukraine, Russia has focused its attacks solely on Ukraine's electrical grid.

A lesson that must reverberate throughout the global community is that Ukraine's civilian resilience and dedication at all costs to the war effort has kept it in the war and has shaped an entirely new army. In contrast, Russia has stuck with World War II tactics and has failed due to poor leadership, poor training, and a poor culture within its military. Russia lacks a professional noncommissioned officer corps that is the backbone of Western armies, necessary in peacetime, and critical during combat operations. In 1997, the US Army Chief of Staff, General Eric K. Shinseki, met with his Russian counterpart, General Alexey Maslov, and after initial pleasantries were exchanged, General Maslov told General Shinseki that he had the best tank in the world. General Shinseki paused momentarily and then replied, "No, General, not only do I have the best tank in the world, but I have the best noncommissioned officers corps. The latter alone will guarantee victory for us no matter where we fight or who the enemy might be." The Russians have lost many of their generals simply because the Ukrainians adapted their warfighting and surprised the Russians. When the generals move forward to inspire and direct combat operations, they are often wounded or killed because the Russian war system does not reward initiative or innovation and requires direction even on the front lines of battle. The Ukrainians have proved to be nimble, agile, and willing to adapt and learn on the go. This alone was the saving grace for Ukraine and not all the technologically superior weapon systems given to it. In the early stages of the war, the Ukrainians received Western night vision goggles, which they immediately used with snipers hiding in the woods. The Russians lacked this equipment, so they remained on the roads, making it much easier for the Ukrainian snipers to distinguish friends from foes. Another surprise was the Western medical kits that were given to Ukraine and once again, they were immediately used and saved countless wounded Ukrainian soldiers. The Russians, on the other hand, lacked these items, and many wounded bled to death awaiting medical care that never came. In what the US Army would refer to as psychological operations, Ukrainian civilian groups have been organized and have been calling Russian families to share ongoing stories from the front that differed considerably from what the Russian media was dispensing back in Russia. Ukrainian soldiers even give captured Russian soldiers a cell phone so they can call their families back in Russia to share the news of their capture and ethical treatment and to provide updates on the war. If missiles attack schools, the schools are relocated, and the education of the youth continues. Trains have been running throughout the war, even with sporadic attacks on stations, primarily due to railway workers communicating with one another on when to move, who has the right of way, and which side of the tracks their trains should be on. These two simple examples depict a society that is all-in on its intent to outlast the Russian invaders. The younger generation is tech-savvy, similar in many ways to every other nation, with one exception: they are fighting for their nation's survival, are highly adaptive, and are willing to seize upon mistakes and opportunities that the Russians provide. They have figured out how to build cheap UAS, affix explosives, and release mechanisms. This example of innovation has helped keep Ukraine in the war. They created a solution and immediately tested it against the Russians on the battlefield. Unlike modern armies, there is no bureaucracy nor a long, systematic development process to develop systems needed today on the battlefield.

The US and NATO nations have answered the call by Ukrainian President Zelenski to provide the Ukrainian military with weapons systems and ammunition to battle the Russians and win back their territory. This includes the transfer of 28 long-range multiple-launch rocket systems, including the truck-based M142 High Mobility Artillery Rocket System (HIMARS) and the M270 Multiple Launch Rocket System (MLRS). These rocket systems were designed to wreak havoc far behind the front lines, and that is precisely what they have done with Russian lines of communication, ammunition depots, air defense systems, and headquarters. In concert with modern weapons systems, the Ukrainians have become experts at using UAS to conduct battlefield surveillance and locate targets, and with high-speed command and control, they can execute attacks on those targets very quickly. Without a similar capability, Russia cannot identify or attack the rocket systems that are destroying their forces. Future wars will still be won or lost via traditional military equipment supported by new technologies. Ukraine is flying hundreds of UAS a day to surveil Russian equipment and positions and to target them with astonishing speed. UAS seek out Russian artillery positions while 3-D mapping allows Ukrainian forward observers to pinpoint targets to their headquarters, which in turn, launch artillery, UAS, ballistic or cruise missiles toward them.

Forensics

In early 2017 in eastern Ukraine, Ukrainian military forces shot down a six-foot-long Russian UAS that had been conducting surveillance over them (Whalen, 2022). When researchers examined its contents, they were surprised to find electronic components that Western companies had entirely manufactured: "The engine came from a German company that supplies model-airplane hobbyists; US suppliers made computer chips for navigation and wireless communication; a British company provided a motion-sensing chip; other components were manufactured in Switzerland and South Korea" (Whalen, 2022). Later, the London-based Conflict Armament Research Group (CAR) verified this, examining the components on a trip to Ukraine. It is unlikely that trade sanctions would be effective since many components are rather mundane ones that would be hard to track, are required in small numbers, and are used in many products inside Russia. In the 1990s, the Soviet Union had a small number of semiconductor factories producing electronic chips, primarily for use in military equipment. The dissolution of the Soviet Union led to an extended period of internal turmoil that obstructed the advancement of various technological industries and manufacturing. "The microelectronics industry was completely decimated in the 1990s. It became far easier to simply import these technologies, which were widely available in the global market" (Whalen, 2022). Most electronics found in Russian military equipment come from Western goods stripped of extraneous components, and the electronic parts are reused in military equipment. CAR estimates that between 2013 and 2016, Western suppliers were more dominant in the chip industry, but since then, many chips have now been produced in China and Taiwan. Russian companies are believed to send computer chip designs to the Taiwan Semiconductor Manufacturing Co., the world's largest chip foundry, for fabrication (Whalen, 2022). China would most certainly ignore any sanctions levied against the export of computer chips to Russia. According to the United Nations trade data, Russia relies on "Asian and Western countries to supply most of its consumer electronics and computer chips, estimated in 2020 to be greater than $38 billion" (Whalen, 2022).

Orlan-10 photograph courtesy of J. Whalen, 2022, https://www.washington
post.com/technology/2022/02/11/russian-military-drones-ukraine/

During its existence, the Soviet Union reportedly had many espionage successes stealing technological secrets from US companies and government agencies. Today the vulnerabilities of Western companies and government secrets continue. The dissections of Russian UAS have revealed their capabilities and limitations. Still, they have also resulted in uncovering the existence of an elastic, sanctions-evading supply chain that stretches into Western countries and Asia. In 2015, several Russian agents were convicted of multiple US federal charges of illegally using a fictitious Texas-based company to export high-tech chips to Russian military and intelligence agencies. There is little that the West can do to stem the supply of electronics to Russia for use in their weapon systems or to better protect its secrets from Russian and now Chinese spies.

In 2017, a Russian UAS was downed in eastern Ukraine and investigators found components made in the Western and Japanese camera parts inside. At the behest of the Ukrainian government, experts from CAR conducted a forensics examination of the UAS and attempted to run down the supplier of its non-Russian components. The team disassembled and photographed every component of the UAS, including serial numbers to identify where the parts originated. The following is an excerpt from this forensics search:

- "One motion-sensing chip was manufactured by the British company Silicon Sensing Systems, which makes components for drones, car navigation systems, and industrial machinery. Silicon Sensing Systems stated that it had sold the chip in August 2012 to a Russian civilian electronics distributor, sending it through UPS in a package with 50-odd components.
- The Russian distributor stated that it later sold the chip to a Russian entity called the Professional Association of Designers of Data Processing Systems (ANO PO KSI), which it said purchased such items for educational institutions in Russia.
- ANO PO KSI was added to a sanctions list by the United States in 2016 for allegedly aiding Russian military intelligence.
- Silicon Sensing Systems said it vigorously complies with all export control laws and policies everywhere it does business and that these components were sold in 2012 to a commercial company that was not on an embargo list at that time.
- We have ceased doing business with that company and any related entities" (Whalen, 2022).

In another case involving another Russian UAS downed in 2017, components included a wireless communications component that was manufactured inside the US.

- "The UAS also contained U.S.-made navigation and wireless communication components.
- One of the suppliers, Digi International, stated that it sold the wireless communications component to a U.S.-based distributor in March 2012" (Whalen, 2022).

In yet another downed 2013 Russian UAS, US company Maxim Integrated noted that it had manufactured a navigation component that it shipped to its distributors.

- "Other components led to companies in Switzerland and the UK that stated that they could not track the chain of suppliers that had handled their components.
- The engine that powered the UAS originated from 3W Modellmotoren Weinhold in Germany, which specialized in model airplane parts.
- 3W Modellmotoren Weinhold stated that it had sold the engine to the World Logistic Group, a Czech Republic company that ceased operations in 2018.
- The company was founded in the spa town of Karlovy Vary in 2008 by two residents of Moscow, according to Czech business registration documents.
- From 2012 to 2014, a third Moscow-area resident served as a director of the company, according to those documents.
- Researchers found that this person was also an advisory council member to the Main Directorate of Public Security for Moscow's regional government.
- The directorate was established to implement state policy in public and economic security, according to the website of Moscow's regional government" (Whalen, 2022).

Lithuania discovered an identical model that crashed on its territory in 2016. It also contained Western components and Russian software. According to Lithuanian authorities, the case demonstrated that "Russia uses UAS for intelligence collection not only in conflict zones but also in peacetime in neighboring NATO countries" (Whalen, 2022).

According to CAR weapons experts, the engines manufactured in Iran for use in its Shahed 136 UAS are based on technology stolen from Germany over seventeen years ago (Baker, 2023). Russia has redirected 24 SU-35 fighter aircraft destined for Egypt to Iran as partial payment in return for its hundreds of UAS. Egypt feared

US-led economic sanctions and wanted to back out of the deal fighter aircraft deal. Russia also included captured Western military systems to replenish its contract with the Iranians to gain UAS. It has used the UAS to target critical infrastructure within Ukraine, primarily electrical power systems and networks. After closely examining downed Iranian-manufactured UAS, the engine was traced to an Iranian company, Mado, and bore an uncanny resemblance to the German Limbach L-550 engines produced decades prior. CAR also determined that other UAS models produced by Mado are being mass-produced based on other European technologies and UAS models.

According to reports from the company, the Russian weapons manufacturer Uralvagonzavod is developing an AI-enhanced USS main battle tank called the Sturm based on the T-72B3 tank chassis (Shestopalov, 2023). The Sturm is being created with four different armaments. The first is equipped with a 125 millemeter cannon, the second has missiles or a rocket-propelled infantry flamethrower with a thermobaric Shmel-M warhead, the third is equipped with two 30mm guns, and the fourth would operate in conjunction, possibly as a support vehicle, with the Russian TOS-1A thermobaric rocket launcher system (Shestopalov, 2023).

Counter-UAS

The much-feared Russian military has failed in the Russia-Ukraine War across the board in so many different areas. According to David Axe, a Forbes military analyst, Russia attempted "too many attacks along too many sectors, which thinned out Russia's best battalions; too few infantry units to screen the tanks; inflexible air support; artillery batteries that bombarded too many empty grid squares; and perhaps most importantly: inadequate logistics for what would become a long, grinding war" (Axe, 2023). Although it was deployed after major fighting units moved into Ukraine, the Russian electronic warfare units suppressed Ukraine's main asset for collecting and reporting on battlefield intelligence; its UAS fleet. Ukrainian Air Force pilots "had their air-to-ground and air-to-air communications jammed, their navigation equipment and radars would be knocked out" (Axe, 2023). According to the Organization for Security and Co-operation in Europe (OSCE), the Russian army deployed a significant number of electronic warfare systems in Russian-occupied eastern Ukraine. They included the "TORN and SB-636 Svet-KU signals-intelligence systems that could pinpoint Ukrainian units by tracing their radio signals, RB-341V Leer-3s that teamed an Orlan-10 UAS carrying cellular-jamming payloads with a command post on a KamAZ-5350 truck, R-934B Sinitsa radio-jammers and R-330Zh Zhitels that block satellite links" (Axe, 2023). According to OSCE, the Russian electronic warfare force was so potent that it also jammed Russian UAS and aircraft (Axe, 2023).

The age of the camera video recordings has long been with us and now there are multiple unconfirmed reports based on poor video footage, apparently taken from a Ukrainian UAS, that shows a Russian DJI Mavic quadcopter drone with the letter Z painted on it being rammed by an unidentified Ukrainian UAS. There was a sudden change in camera angle from the Ukrainian UAS, which could have been caused by it flying into and ramming the Russian DJI. After the collision, the Russian UAS retreated from the Ukrainian UAS. In October, a Ukrainian filmed a similar battle between Russian and Ukrainian DJI Mavic quadcopters where the Ukrainian drone flew to the Russian UAS and rammed it thereby damaging its rotors, resulting in it crashing to the ground. The reported video footage could be genuine or merely Ukrainian propaganda, but according to Ukrainian Ministry of Defense spokesmen, Ukraine UAS operators are targeting and attacking Russian UAS in the air when they encounter them. Regardless of fact or fiction, it is a "glimpse of future aerial combat that could become the mainstay of wars of the future" (McFadden, 2022).

The German-made Gepard system, twin anti-aircraft cannons atop a tracked vehicle, has provided Ukraine with a mobile air defense system vital in destroying Russian UAS and missiles aimed at power stations and electrical generating plants.

McCleary, P. (2023). The little-known weapon knocking down Iranian drones over Kyiv. https://www.politico .com/news/2023/01/04/weapon-iranian-drones-ukraine-00076442

Gepard can launch two streams of 35mm airburst rounds into the sky to hit its targets. Due to its age and limited supplies of its 35mm ammunition, German manufacturer Rheinmetall said it would open a new production line to manufacture this rare ammunition. This example "highlights a lack of domestic industrial capacity that has increasingly alarmed Western governments since the start of the Russia-Ukraine War" (McCleary, 2023). The US and its European allies are also identifying weapons that could be pulled out of stockpiles and sent to Ukraine. This review includes the US-made Hawk air defense system, which would require excess launchers currently stockpiled in Spain. The US replaced the highly effective Hawk air defense system with the Patriot air defense system back in the late 1970s.

Sanction busting

Russia has been and continues to buy a great deal of technology from the West that it has used for its war against Ukraine. It is now using China as a middleman to avoid export controls. Most of the imported critical components in Russian weapons are made by US companies. According to Russian trade data, Russia received 64 percent of its dual-use goods, parts that can be used for civilian and military products, from US companies between March and December 2022 (Bush, 2023). Russia also bought technology from 155 companies in Europe, Asia, and the Middle East. Ukrainian authorities found that 66 percent of the foreign critical components in Russian weapons used in Ukraine were made by US companies (Bush, 2023). According to a think tank in Kyiv, Russia imported $20.3 billion worth of parts for military equipment from March to December of last year (Bush, 2023). This is only a tiny portion prior to Russia's invasion of Ukraine in February 2022. This demonstrates that some private companies are not following trade rules, Western governments are having trouble enforcing them, and dual-use technology is nearly impossible to track in the global market.

The Russian Orlan 10 UAS has been used extensively in Ukraine and has been instrumental in Russian battlefield surveillance and directing artillery fire on exposed Ukrainian positions and equipment. A 2022 investigation by Reuters and iStories, a Russian media outlet, and the Royal United Services Institute (RUSI), a defense think tank in London, uncovered a logistics trail that stretches from Western countries to the Special Technology Centre in Saint Petersburg, Russia. In 2017, President Obama initiated sanctions against the Centre due to their perceived collaboration with Russian intelligence agencies to meddle in the 2016 US presidential election. The sanctions "barred any American citizen, resident, or US company from supplying anything that might end up with the Special Technology Centre. In 2022, the US government tightened those restrictions by blocking all sales of any American products for any military end user. It effectively blocked all sales to Russia of high-technology items like microchips, communications, and navigation equipment" (Tamman & Zholobova, 2022). The transfer of US technologies and parts should have been halted, but they have continued, as has the production of the Orlan UAS for the Russian frontline soldiers. Russian manufacturers continue to receive Western components despite sanctions for the Orlan-10 drone via intermediaries in the US, China, and Russia. US-supplied parts are often purchased and shipped to one or more third-party entities, intent on circumventing sanctions and, thus, making huge profits. One such third-party intermediary to the Russian Special Technology Center has been Hong Kong-based exporter, "Asia Pacific Links Ltd., which, according to Russian customs and financial records, provided millions of dollars in components, namely microchips from US manufacturers and model aircraft engines made by a Japanese company, Saito Seisakusho, that are used in the Orlan 10" (Tamman & Zholobova, 2022). US sanctions have resulted in problems for Asia Pacific Links Ltd., but it considers it a badge of honor, "Sanctions were imposed on us by one of the most powerful countries in the world, we should be proud of this" (Tamman & Zholobova, 2023). According to a joint investigation by Vazhnye Istorii, Important Stories, Reuters, and RUSI, Russian-made Orlan-10 UAS are still being assembled with components from US-based companies Altera and Xilinx, Texas Instruments, Microchip Technology, Analog Devices, Linear Technology, European STMicroelectronics and NXP Semiconductors, and Japanese companies Renesas Electronics and Saito Seisakusho" (Pravda, 2022). None of the components manufactured by those companies should reach Russia, yet they are. A long line of willing intermediaries has grown rich circumventing sanctions supplying Western components to Russia's defense industry.

Igor Kazhdan is a forty-one-year-old US-Russian citizen who lives in Florida and owns IK Tech, a U.S.-based company that sold millions in electronic components to Russia dating back to 2018. One of the items that his company sold were 1,000 American-made circuit boards, valued at $274,000 each, manufactured by California Gumstix, and banned for export to Russia by federal law (Tamman & Zholobova, 2022). Forensics by Ukrainian parties have shown the internal components of an Orlan-10 is a Gumstix circuit board. Kazhdan also sold sophisticated amplifiers made by US-based Qorvo, banned for export to Russia, which are often used in radar, communications, and radio equipment and have also been found inside downed Orlan-10 UAS (Tamman & Zholobova, 2022). Kazhdan was eventually arrested and pled guilty to two lesser charges to which a federal judge sentenced him to three years of probation, fined him

$200, and ordered him to forfeit about $7,000, a far cry from the forty years he could have been sentenced to if convicted on all counts in his original indictment (Tamman & Zholobova, 2022). With sweet punishment deals such as this, it is no wonder that there is no shortage of people standing ready to make huge profits skirting US export laws and providing much-needed supplies to Russia, or possibly any other country willing to pay. It should also come as no surprise that the smuggling of components will continue unabetted. When asked about Kazhdan's light sentence, the US Department of Justice declined to comment on the case.

Three Russian nationals were arrested in New York in November 2023 on charges of wire fraud, smuggling, and conspiracy to violate the Export Control Reform Act, by shipping an estimated $10 million in semiconductors, integrated circuits, and other electronic components with dual civilian and military uses to sanctioned entities in Russia (Vigliarolo, 2023). Nikolay Goltsev, Salimdzhon Nasriddinov, and Kristina Puzyreva sent three hundred shipments through front companies to Turkey, Hong Kong, India, China, and the UAE before the shipments were rerouted to Russia (Vigliarolo, 2023). The components were discovered inside Russian UAS after Ukrainian air defense forces shot them down, and forensics experts examined the wreckage.

The US issued new sanctions in January 2023, targeting the suppliers of Iranian UAS that Russia is using to target civilian infrastructure in Ukraine. The US Treasury Department stated that "it imposed sanctions on six executives and board members of Qods Aviation Industries, also known as Light Airplanes Design and Manufacturing Industries, which has been under U.S. sanctions since 2013" (Psaledakis & Mohammed, 2023). Treasury Secretary Janet Yellen also directed sanctions against the director of Iran's Aerospace Industries Organization, which is responsible for overseeing Iran's ballistic missile programs. This action freezes any US assets and bars Americans from dealing with them. A single Iranian UAS downed in Ukraine late in 2022 was found to contain parts manufactured primarily by US companies. Out of a total of "52 components removed from the Iranian Shahed-136 UAS, 40 appear to have been manufactured by 13 different American companies, and the remaining 12 components were manufactured by companies in Canada, Switzerland, Japan, Taiwan, and China" (Bertrand, 2023). This is proof that despite sanctions, Russia and Iran are still finding abundant commercially available technology. The Iranian company that built the downed UAS, Iran Aircraft Manufacturing Industries Corporation, has been under US sanctions since 2008 (Bertrand, 2023). The global microelectronics industry relies on third-party distributors and resellers, and since computer chips are small, light-weight, easily transferrable, and shipped, they are worth a lot of money, which entices smugglers to the game. It is inconceivable that the smuggling of sanctions parts will be stopped, but Western governments can increase their enforcement efforts and make it much more complex and expensive for nefarious parties to get their hands on these highly sought-after electronic components.

Asia Pacific Links Ltd., a Hong Kong-based exporter, has been a critical parts supplier for Russia's UAS program. The company, owned by Russian expatriate Anton Trofimov, has never dealt directly with Russia but has shipped millions of dollars' worth of microchips from US manufacturers through intermediaries. One of its main customers is SMT iLogic, a St. Petersburg–based importer that works closely with the Special Technology Centre, the UAS manufacturer. SMT iLogic and Asia Pacific Links have significantly increased trade since Russia invaded Ukraine in February 2022. According to Russian customs records, Asia Pacific Links sent parts worth about $5.2 million to SMT iLogic between March and September 2022, compared to about $2.3 million in the same period of 2021 (Grey et al., 2022). The records also reveal that some parts were made by US tech firms such as Analog Devices, Texas Instruments, and Xilinx (Grey et al., 2022). One of the parts that Asia Pacific Links supplied to SMT iLogic was a model aircraft engine made by a Japanese company, Saito Seisakusho. This engine is being used in the Russian Orlan 10 UAS, which was confirmed after one was shot down and recovered in Ukraine. According to customs records, SMT iLogic is a major importer of electronic products in Russia, with about $70 million worth of imports since 2017 (Grey et al., 2022). The company's main partner is the Special Technology Centre, which makes UAS for the Russian Ministry of Defense, which earned $99 million from the ministry between February and August 2022 (Grey et al., 2022).

According to the former Pentagon official Gregory Allen, who now serves as director of the Artificial Intelligence Governance Project at the Center for Strategic and International Studies, "Nobody has really thought about investing more in agencies like the Bureau of Industry Security, which were really sleepy parts of the DC national security establishment for a few decades. The branch of the Commerce Department that deals primarily with export controls enforcement I snow suddenly, they're at the forefront of national security technology competition, and they're not being resourced remotely in that vein" (Bertrand, 2023). Damien Spleeters, the deputy director of operations at Conflict Armament Research, states that "sanctions will only be effective if governments continue to monitor what parts are being used and how they got there. Iran and Russia are going to try to go around those sanctions and will try to change their acquisition channels, and that's precisely what we want to focus on: getting in the field and opening those systems, tracing the components, and monitoring for changes" (Bertrand, 2023).

In February 2023, the Ukrainian military, and CAR examined an unexploded Shahed 136 UAS that had crashed in the Ukrainian city of Odesa. The Russians made a deal with Iran to purchase unmanned aerial systems, which it has used to target Ukraine's critical infrastructure, namely its power grid and distribution centers. According to the examination of the unexploded Shahed 136:

> The warheads were hurriedly modified with layers of tiny metal particles that, upon contact, scatter across a wide radius. There are 18 smaller charges around the warhead's diameter in addition to the fragments, which, when melted by the blast, can pierce armor and have a 360-degree explosive impact. Those components effectively increase the warhead's capacity to rip apart targets, including power plants, distribution grids,

transmission lines, and sizable, high-power trans-
formers. Additionally, they significantly compli-
cate repair operations. (Tripathi, 2023)

UAS targeting military systems such as tanks would use a warhead with a shaped charge to penetrate the armored vehicles. The
Iranian Shahed 136 has been successful in hitting the Ukrainian
power sector faster than the nation's power provider, Ukrenergo, can
affect repairs. The slow-moving, propeller-driven range of the Shahed
136 version has been estimated by military analysts as between 600
to 1,200 miles. The Ukrainian military claims a 60 to 80 percent
success rate in shooting them down. The cost of shooting down
these Iranian-made UAS by Ukraine can undoubtedly be the cost of
manufacturing them. The Shahed 136 UAS cost between $20,000
to $50,000 to manufacture, yet the counter systems used by Ukraine
cost three times or more than that (Jankowicz, 2023).

The larger Shahed 131 was reportedly used to conduct an attack
inside Saudi Arabia with an estimated range of 420-600 miles. The
attack against the Saudi oil facilities at Abqaiq and

Khurais occurred in September 2019 and were damaged by
UAS and cruise missiles. Houthi insurgents allegedly launched the
attacks in retaliation for Saudi Arabia's intervention in the Yemen
civil war (Rubin, 2023). A few days later, debris from the attacking UAS was exhibited in a press conference convened by the Saudi
Armed Forces and was identified later as an Iranian Shahed 131
UAS. On 21 July 2021, the oil tanker *Mercer Street* was struck by an
unknown unmanned aerial system, killing two (Rubin, 2023). The
ship was hit near the Straits of Hormuz and forensic examination of
the unmanned aerial system debris found on Mercer Street revealed
that it was identical to the UAS used in the September 2019 Saudi
attack.

Iran's initial batch of UAS to Russia had an unusually high
number of malfunctions. Russia did not have a large inventory of
UAS and exhausted them quickly; thus, it was forced to seek replacement UAS from one of its closest partners.

Image of an Iranian Shahed 136 drone, courtesy of Tripathi, A., 2023, Iran 'Modified' Shahed-136 Kamikaze Drones For Russia To Cause Max Destruction On Ukrainian Infra. https://world-defence.com/ukraine-war-iran-modified-shahed-136-kamikaze-drones-for-russia-to-cause-max-destruction-on-ukrainian-infra/

Russia did not grasp the significance of UAS at the onset of the war but quickly changed this view thanks primarily to Ukraine's game-changing use of them against Russian troops and military equipment. At the start of the war, Russia had an arsenal of UAS for service in the Russian Ground Forces as a component of its reconnaissance-strike complex doctrine, where UAS would quickly conduct battlefield reconnaissance to locate targets for subsequent artillery strikes (Mizokami, 2022). The prewar Russian arsenal included "Orlan-10, Orlan-30, and Zastava, and larger medium-altitude, long-endurance (MALE) reconnaissance UAS such as Forpost-R and Orion, all designed for low-altitude intelligence gathering operations, surveillance, and reconnaissance missions (Mizokami, 2022). An independent analysis of battlefield losses has the Russian losses at "5,362 tanks, aircraft, armored vehicles, howitzers, trucks, and UAS. According to Ukraine's high command, Russia has lost even more, including over 847 UAS (Mizokami, 2022). UAS are lucky to survive a mission, with Ukraine UAS lasting no more than three missions. UAS are sent to hazardous areas where they face engagement by SHORAD units, small arms fire by soldiers on the ground shooting their rifles at them, operator error, and mechanical issues. Due to their relatively cheap costs and hazardous missions, these UAS are expected to have a high percentage of losses. Russian agreement to purchase Iranian-manufactured Mohajer-6, Shahed 129, and Shahed

191 UAS is ironic since, for many decades, Iran had purchased its weapon systems from Russia (Mizokami, 2022).

The use and evolution of unmanned systems in the Russia-Ukraine War have accelerated the incorporation of AI into these systems, allowing them to function as an autonomous entity capable of selecting and engaging targets without human control or approval. This will be a much-anticipated revolution in modern warfare, as significant as hand grenades, machine guns, and aircraft years ago. According to Ukraine's digital transformation minister, Mykhailo Federov, this is a logical next step and Ukraine has been conducting a great deal of research and development in this arena (Bajak & Arhirova, 2023). Currently, loitering munitions such as the US-manufactured Switchblade, Israeli Harpy, and Polish Warmate require human beings to select targets, after which AI completes the attack. As we encounter AI against AI-enhanced military equipment, a much faster path would be to eliminate the human from the kill chain, and program the parameters of the targets and authorizations to engage into the AI algorithm and launch. These autonomous platforms would loiter for as long as their fuel supply would allow, all while looking for targets. Prior to fuel exhaustion, these platforms could select targets far down on their list of high-value targets to explode and spread terror to ground troops. As previously mentioned, there was a 2020 United Nations report that stated that during "Libya's internecine conflict, Turkish-made, and autonomous Kargu-2 UAS killed an unspecified number of combatants without human operator approval (Bajak & Arhirov, 2023).

According to President Putin, AI is a priority for Russia. In a 2017 speech, he stated, "Whoever dominates that technology will rule the world. He also expressed confidence in the Russian arms industry's ability to embed AI in war machines, stressing that the most effective weapons systems are those that operate quickly and practically in an automatic mode" (Menendez, 2017). He later remarked in that same year during a televised session with an engineering student, that "when one party's drones are destroyed by drones of another, it will have no other choice but to surrender" (Menedez, 2023). It is interesting to note that a University of California-Berkeley professor and top AI researcher, Stuart Russell, and his colleagues in the mid-2010s, agreed that graduate students could, in a single term, produce an autonomous drone "capable of finding and killing an individual, let's say, inside a building" (Menedez 2023).

Russia has provided military equipment to Iran for years, so imagine the irony for Russia to have to ask Iran for UAS to replace its depleted stocks. It is unlikely that any other nations would dare to sell UAS to Russia for fear of Western sanctions and political pressure. For many years, Iran has provided military equipment, including UAS, training, and funding to its proxy militias in Bahrain, Iraq, Lebanon, Syria, and Yemen primarily via its elite Quds Force, the external operations branch of the Revolutionary Guards (U.S. Institute of Peace, 2023). The Iran-supported Houthi rebels in Yemen have been launching UAS attacks against Saudi Arabia, no doubt in retaliation for its support to the recognized government of Yemen, for over seven years. Iran also gained experience and possibly technical knowledge which it could put to good use to improve its UAS exports to Russia, from its successful 2019 attack on the Saudi Arabia oilfield and refinery located in Abqaiq-Khurais. It is interesting to note that Russia's pursuit of Iranian UAS proves that they have had a change of heart and now believe from experience inside of Ukraine, that UAS have become an essential military capability in modern warfare. It also highlights that the Russian industrial complex, which has struggled to build and even repair damaged tanks and other military equipment, is incapable of building UAS to meet increased Russian demand on the battlefield. In the second year of the Russia-Ukraine War, Ukraine has improved the quantity and quality of its air defense force by acquiring US Stinger, SHORAD systems, PATRIOT, NATO NASAMS, and IRIS-T air and missile defense systems. Clearly, the international sanctions have had a detrimental effect on Russia's ability to gain the components necessary to build additional UAS platforms. All these factors reflect long-term, economic problems for Russia and the poor state of its industrial base. As stated previously, the Russian invasion of Ukraine has had the opposite effect on NATO than Putin envisioned at the onset. European Union leaders agreed to bolster European economic resilience, increase military budgets and capabilities, and drastically reduce their reliance on Russian oil. The Russian Ruble has "lost half its value, inflation is soaring, Moscow's stock exchange is closed, and many international companies have left the country; as a result, Russia's economy is expected to shrink by at least 15 percent in 2022" (Borrell, 2022). This weakened and isolated Russia is now more than ever dependent on China and its recovery from the failed Special Security Operation inside Ukraine will take decades.

Cargo-delivery UAS in Ukraine

At the beginning of the war, there were calls for UAS to resupply besieged Ukrainian forces and deliver humanitarian aid. The truth is that UAS delivery systems were not ready for this mission and may still not be. According to a former pilot and UAS entrepreneur, "Limited drone deliveries might hold symbolic value, provide valuable learning, and occasionally meet legitimate needs, but are unlikely to meaningfully impact Ukrainian logistical requirements (Jacobsen, 2022). In the future, C-UAS delivering cargo will undoubtedly play an important role in combat. For now, the global community has simply not invested the time, technologies, or resources to make UAS deliveries a reality. For many of those same countries, the revolutionary use of UAS in combat caught them by surprise. Now that it is nearly two years into the war, we may see governments make the necessary long-term investments to create military-grade UAS delivery systems capable of adding to battlefield logistics. should take the lesson and make long-term investments in last-mile battlefield logistics capabilities. Less than a decade ago, Chinese company DJI took the world by storm with the release of their Phantom quadcopter. What differentiated it from its competition was primarily its advanced software and "sophisticated autopilot allowing operators to fly the drone out of the box, with no training or experience" (Hambling, 2022). The onboard camera allowed companies and government agencies to use UAS to conduct aerial inspections of facilities, bridges, buildings, monuments, and critical infrastructure, faster and cheaper than a helicopter could. Within a short period of time, DJI had established itself as the premier UAS manufacturer with 76 percent of the global sales for sUAS (Slotta, 2022). In 2021, one of the largest arsenals of commercial sUAS, many of those being DJI, was estimated to number 813 and was owned and operated by the US Department of Interior. By 2017, ISIS had developed an aiming mechanism and the means for attaching explosives and routinely dropped aerial improvised munitions on US and coalition forces in Iraq and Syria.

The explosives were usually captured US-made 40mm grenades that had been modified with the addition of tail fins and a different fuse (Hambling, 2022). ISIS took advantage of social media and posted hundreds of video clips of UAS releasing explosives personnel in the open and vehicles in Iraq and Syria. This improvised form of warfare spread quickly to other nations, such as Afghanistan, Myanmar, and Mexico, predominantly by the drug cartels for use against one another, the Central African Republic, and in Ukraine, and Russia.

Until recently, little was known about a "non-governmental organization of volunteers and IT specialists" in Ukraine who make up the Aerorozvidka, or Ukraine's Aerial Reconnaissance Force, which has had great success modifying sUAS with explosives and turning them into potent tank killing machines (Hambling, 2022). Their TTPs, mechanisms, and attack methods will likely be shared and mimicked globally. Aerorozvidka was created during the 2014 Russia-Ukraine War to provide surveillance of Russian forces for Ukraine's military forces. They began by using COTS UAS, which they modified to drop explosives on Russian soldiers and vehicles. Still, they concluded that a much larger UAS platform made from existing Ukrainian commercial components was required to drop the necessary number of explosives to pierce the tops of tanks and stop them in their tracks.

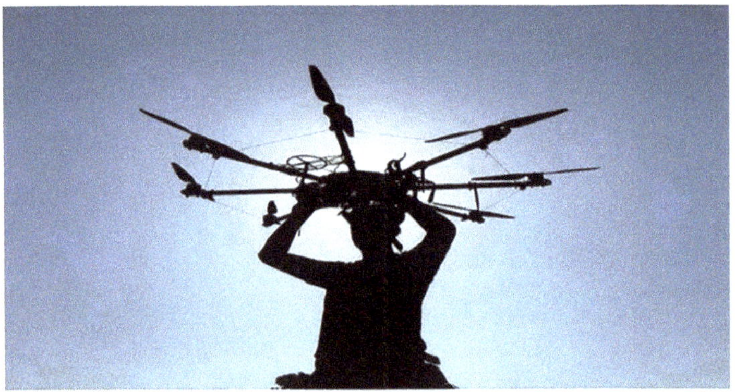

Image courtesy of Hambling, D. (2022). https://www.-forbes-com.cdn.ampproject.org/c/s/www. forbes.com/sites/ david hambling/2022/06/01/how-ukraine-perfected-the-small-anti-tank-drone/amp/

The result of their efforts was the R18 octocopter, which has eight rotor blades, a flight time of forty minutes, a lift capacity of eleven pounds, and is fitted with a thermal imager that can discern heat sources from engines in the dark and hidden behind vegetation (Hambling, 2022). The R18 usually takes off with three RKG-1600 bombs, each weighing 2.2 pounds and adapted to existing stocks of seventy-year-old Soviet Union anti-tank grenades (Hambling, 2022). The RKG-1600s have been retrofitted with plastic fins, which allow them to be dropped accurately from a UAS 300 feet in the air. The warheads are powerful enough to breach over 200mm of steel, easily penetrating the top of a tank's turret (Hambling, 2022).

A proven technique for the Aerorozvidka is to fly their R18 above a target and release the first of their three RKG-1600 bombs.

The operator will watch via the onboard camera to see where it lands, then adjust the location of the R18 above the target. With these silent platforms silently flying 300 feet above the target, the targets may not know that they are under attack or if they do, they will not know where the R18 is. It is normal for Russian crews not to know that they are under attack when one considers that their vehicle engine is running, so it will easily drown out the sound of a small explosive hitting outside of it. Normally, Russian soldiers will fire indiscriminating into the air with machine guns, but the odds of hitting a UAS flying 300 feet in the air are minuscule. The second or third bomb is likely to hit its designated target in the specific area in which it must destroy the target. Unlike their NATO counterparts, "Russian tanks do not store ammunition separately; thus, any penetrating hit on the crew compartment will trigger a catastrophic ammunition explosion (Hambling, 2022). This technique provides a high degree of cost-effectiveness from three low-cost, RKG-1600 dumb bombs. Aerorozvidka generally attacks at night, which allows for a rapid adjustment of bombs dropped onto targets, thereby increasing its effectiveness while instilling fear in those Russian soldiers operating in the dark and in the vehicles that have been targeted.

According to Aerorozvidka, the R18 UAS cost $20,000 apiece and can stop all Russian tanks, even those equipped with layers of reactive armor, active protection systems to intercept incoming projectiles before they hit the tank, and even bolt-on cage armor designed to counter-UAS attacks (Hambling, 2022). Aerorozvidka claims that as the war continues, they refine their tactics, techniques, and munitions, thereby keeping the R18s highly lethal and effective in their terror campaign against a poorly led Russian army. In addition to the RKG-1600 bombs, other munitions have been developed from the warheads of old, Soviet-era weapons in abundance in Ukraine. In June, Ukraine sent a low-cost Punisher UAS that cost $10,000 to drop explosives on and destroy "one Tor-M2 air defense system worth $25 million, a URAL cargo truck, and two BM-21 Grad multiple rocket launchers, and an ammunition depot" (Panasovskyi, 2023f).

Ukraine and its army of volunteers, primarily citizens with experience or education in areas needed desperately by their government, have answered the call to arms. They are using their expertise to build hundreds to thousands of homemade UAs each month. The Ukraine and Russian UAS have truly altered modern warfare as we are witnessing UAS against ground and even UAS aerial engagements. These volunteers have built cheap for less than $2000 each, many under $500, that are deadly, yet easy to build and operate. Recently, a Khrush kamikaze UAS was built and disseminated for use by Ukrainian soldiers located near the front lines. It travels in pieces in a soldier's backpack, but the Khrush can be assembled and armed within three minutes for action. When it is launched, the operator has full control via goggles to surveil the area and to guide the Khrush to a target. The 6-pound explosive payload can destroy any Russian military vehicle (Barnes, 2023).

Saker, a Ukrainian company, started developing AI for small businesses in 2021, especially for crop monitoring using UAS vision systems. However, when Russia invaded Ukraine under the pretext of fighting Nazism, Saker shifted its focus to support the military. Its AI algorithms use machine learning, which means they improve and grow with more data. The company claims that its AI-UAS combination can identify 64 different military vehicles.

Image courtesy of Hambling, https://www.forbes.com/sites/davidhambling/2023/10/17/ukraines-ai-drones-seek-and-attack-russian-forces-without-human-oversight/

The Saker Scout is a drone that can carry almost 7 pounds of explosives and fly over 7 miles. It can also resist Russian jamming by using its AI system to navigate and recognize landmarks. The Saker Scout is part of the Ukrainian "Delta intelligence distribution system that integrates data from drones, satellites, and other sources to create a complete map of the battlefield" (Hambling, 2023). The Saker Scout can autonomously scout an area, locate enemy vehicles, and help with reconnaissance analysis and quick decisions to attack targets. This near-automated process gives an advantage in speed and efficiency, as it does not require human intervention. The Saker Scout can also engage Russian soldiers on its own when communication with its operators is not possible, but only on a small scale. In January, Mykhailo Fedorov, Ukraine's Minister for Digital Transformation and leader of the Army of Drones initiative stated in an interview that "autonomous weapons were a logical and inevitable" next step in drone development, implying a covert but deliberate policy. Fedorov revealed in an official statement on October 6th that the Ukrainian army had just received about 2,000 AI-enabled drones through the Army of Drones initiative. They will help us conduct reconnaissance, adjust artillery fire, and detect even well-hidden Russian targets using AI" (Hambling, 2023). Fedorov said. He did not mention if these drones also had the ability to attack autonomously. If they did not, this feature could be added with a software update. The past worry was that only rogue actors would develop this kind of technology. However, the ethical dilemma becomes more complicated when it is used in a desperate struggle for Ukraine's survival against a vicious invasion. "Ukraine may care less about the potential long-term consequences and more about winning the war, as with cluster bombs. But these weapons will not stay in Ukraine for long. Both sides are likely to be driven by operational pressures toward autonomous weapons while they may start with military targets like tanks and radar, autonomous weapons could soon be used for more indiscriminate methods such as targeting personnel" (Hambling, 2023).

USS

Last year, Ukraine received Tracked Hybrid Modular Infantry Systems or THeMIS fourth-generation USS which have been used to handle tasks that are an immediate threat to soldiers. Now a second shipment will be heading to Ukraine consisting of 14 THeMIS USS, a fifth-generation robotic platform developed by Milrem Robotics and KMW, and the German Ministry of Defense is funding the effort. Half of the THeMIS heading to Ukraine will be configured for casualty evacuation and the second half will be configured for route clearance with payloads. The THeMIS cargo version has been designed to "reduce the cognitive load of soldiers and provide a means to carry and utilize extra gear and firepower. The THeMIS Cargo can also be used to support on-base logistical activities and for last-mile resupply" (Papadopoulos, 2023). With a diesel and electric generator, the THeMIS can carry 2645 pounds and speed up to 12 miles per hour. It has infrared, thermal, and high-dynamic range cameras for vision.

Image courtesy of MILREM Robotics, https://milremrobotics.com/defence/

THeMIS has a self-stabilizing weapon system that can fire accurately at long distances, day or night and it can be armed with different weapons, such as machine guns, grenade launchers, autocannons, or anti-tank missiles (Papadopoulos, 2023). It can also evacuate battlefield casualties and clear routes of explosive devices (Davis, C., 2022). A Russian think tank offered $16,000 to any soldier who captured one so Russia could copy it. According to Jüri Pajuste, director of research and development at Milrem Robotics, "Casualty evacuation and route clearance are two labor-intensive activities that require the engagement of several people who remain in constant threat of enemy fire. Automating these tasks with unmanned vehicles alleviates that danger and allows more soldiers to stay in a safe area or be tasked for more important activities" (Davis, C., 2022). Since February, an estimated 100,000 Ukrainian soldiers have been killed, many because of Russia's indiscriminate use of Improvised Explosive Devices (IED). According to Ruslan Pukhov, the director of Moscow's Centre for Analysis of Strategies and Technologies, "The conflict in Ukraine has demonstrated that modern warfare is unthinkable without the widespread use of unmanned vehicles, we are lagging behind" (Davis, C., 2022).

UAS

Atlas Dynamics and its founder, Ivan Tolchinsky, are helping Ukraine with very capable and relatively cheap UAS. The majority of UAS in use by Ukraine in the war are at the front lines performing "surveillance and reconnaissance missions; guiding indirect fire by artillery, tanks, and mortars; and assessing the results" (Hambling, 2023a). Atlas Dynamic UAS are being built to be immune from jamming and have three sets of rotors with two in the front and one behind, which offers "superior aerodynamics, resulting in a better combination of speed, endurance, and altitude than a quadcopter" (Hambling, 2023a). The Atlas Dynamic UAS has a unique setup referred to as mesh radio, in which each radio is a node that communicates with all other nearby nodes, thereby forming a robust network. If a UAS is unable to communicate directly with the operator but can communicate with another nearby UAS, it will be able to connect to and communicate with the operator. This can occur via multiple hops between nodes, and it currently allows up to five

UAS in a network, but the plan is to build this number to fifty. Tolchinsky plans to evolve his system into an ecosystem, which will include different types of unmanned systems working together in the same mesh. There could be a quadcopter or fixed-wing UAS circling high above, acting as a communications relay with a mix of UAS below, each with different sensors or other capabilities, and even a USS on the ground or atop the sea. Each operator will become a swarm commander, managing a team of unmanned systems, which are autonomous and do work on their own, only bothering their human operators when a command decision is required.

The RAM II kamikaze UAS have proven themselves very effective in the Russia-Ukraine War, especially when targeting Russian air defense systems.

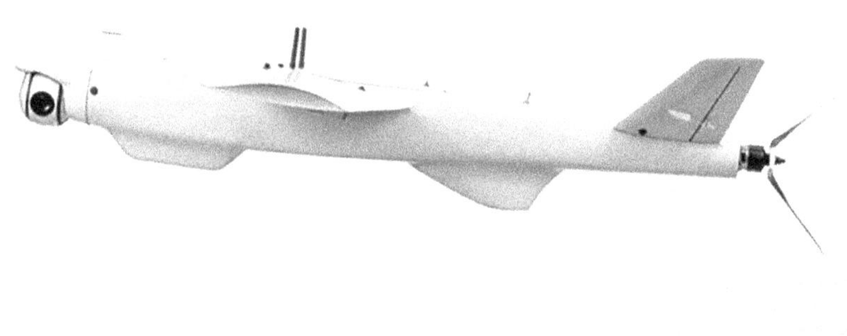

Image provided courtesy of Panasovskyi, M. (2023); https://gagadget.com/en/230470-the-makers-of-the-ukrainian-kamikaze-drone-ram-ii-which-destroyed-tens-of-millions-of-dollars-worth-of-russian-equipment-/

The RAM II is produced in Ukraine and was first developed in 2018 based on a reconnaissance UAS. The RAM II is equipped with a camera that has a 30× optical zoom, a digital image stabilizer, and is capable of viewing objects from miles away and can lock on and engage moving targets. It has a wingspan of nearly 6 feet, weighs 20 pounds, and can stay aloft for 2 hours while carrying a 6-pound explosive, which can be fragmentation, shaped charge, or thermobaric explosive. RAM II can destroy a tank through its turret and all other forms of armored vehicles.

USS

The attention of the Ukraine-Russian War has certainly been on the use and evolving roles of UAS and the use of USS against Russian naval vessels and bridges. Ukraine developed long-range UAS to attack targets in Moscow, though President Zelenski vehemently disputes this, USS has been used to launch military strikes against the Russian navy in the Black Sea (Newdick, 2023a). They have struck infrastructure at the ports of Novorossiysk and Sevastopol and recently have launched missile attacks against Russian air defense and naval headquarters located in Crimea. Many Ukrainian engineers, IT specialists, and aerospace and tank engineers were trained in their craft in the Soviet Union. They are now motivated by their love of their country and hatred of the Russians and are encouraged to use innovation to develop unmanned systems for use by their military for carrying weapons, mainly movable and quickly employable and deployable landmines, machine guns, and soon a 20mm cannon, transport of artillery shells to nearby artillery units, and for reconnaissance (Hunder, 2023). While both Ukraine and Russia have invested in UAS and USS, Ukraine's efforts, courtesy of its innovative populace and small businesses, have outpaced that of Russia. The cost of a USS can be as low as $812, and while its impact on the battlefield has been relatively low, it signals a tactical innovation that could save countless lives as the small, cheap USS traverse the minefields that Russian soldiers laid (Hunder, 2023). The next step would be to evolve from the remote-controlled to the enhanced USS with AI and automation, which would allow them to traverse through contested areas without a human operator. AI is a learning algorithm, and it is only a matter of time before the highly innovative Ukrainians devise a networked environment in which human operators, UAS, and USS collaborate and work together seamlessly on the battlefield. The Ukrainian's tenacity and ingenuity have surprised the Russian military several times. The Ukrainian's tenacity and ingenuity have surprised the Russian military several times. While not the first, certainly the biggest surprise was the Ukrainian attack against the Russian-guided missile cruiser and flagship in the Black Sea, the Moskva. The irony is that the ship's air defense radar, therefore its self-defense systems, was taken offline by an attack by a Turkish-built TB2 UAS. Following this, the Ukrainian military fired two Neptune cruise missiles built on exports that they had received and improved from Russia, and both hit amidship and sank the Moskva. This ability to launch such a coordinated attack caught the Russians by complete surprise and resulted in the remaining Russian naval vessels being moved far out to sea where they would be ineffective or moored in what the Russians believed were two secure seaports. First, the seaport of Sevastopol was attacked by a Ukrainian USS on October 29, then the same Ukrainian USS targeted the Russian Sheskharis oil terminal in the "secure" port of Novorossiysk in November 2022. The USS was apparently an ingenious "homemade device," part jet ski and canoe with two Russian-designed impact fuses that activate when the ship's hull impacts a target and is depressed and trigger the small warhead.

Ukraine surprised the international community and certainly the Russian navy when it unleashed its USS fleet to attack Russian naval vessels and shipping and port facilities in the Black Sea. This was unexpected and embarrassed the Russian military and President Putin. The global community has been caught off guard and very surprised by the success and constant use of UAS on the battlefield. The use of the USS further shocked militaries into action. Ukraine did not spend a great deal of money researching and designing these new USS platforms, rather it took readily available COTS and teamed them with commercially available hobbyist equipment and old but proven Russian detonators and Ukrainian explosives, and at a low cost quickly fashioned all the above into an effective USS. Ukraine's use of USS has served as a wakeup call to navies to relook their capabilities in port security, adding surveillance and rapid-fire weapons to their warships, utilizing USS of their own to provide a frontline against USS or Iranian fast boats as an example, defenses against undersea threats, and mine countermeasures. The undersea realm is only beginning to emerge as an everyday threat to national security and its economy. The recent Nord Stream pipeline attacks are an example of what many countries lack for undersea monitoring and defensive equipment to protect critical undersea infrastructure (Childs, 2023).

One of the support packages that the US sent to Ukraine included USS (boats), manned patrol boats, coastal defense weapons, harpoon missiles, and anti-ship missiles. The number of USS has not been specified, but the US Navy has developed its technology to a point where its vessels are able to "conduct surveillance, targeting of enemy ships, search for mines and network with one another in support of a fleet of manned ships" (Osborn, 2023). The US Navy's multidomain maritime warfare concept is dependent on USSs that are networked with command-and-control nodes undersea, on the surface, on land, and in the air in real-time.

Image courtesy of USNI News, https://news.usni.org/2022/10/11/suspected-ukrainian-explosive-sea-drone-made-from-jet-ski-parts

When teamed with the sixty-two coastal and riverine patrol boats it has sent, it is likely that the US Navy has equipped Ukraine with a networked manned and unmanned fleet that would be able to quickly pass the information on Russian naval vessels to their higher headquarters for targeting, which in turn would send launch orders to the nearest land-based antiship missiles along its coast. Until recently, Novorossiysk was a major naval base, an oil terminal for its

Black Sea fleet, and was considered outside the range of Ukrainian forces (Sutton, 2022c). The Ukrainians have demonstrated an ability to strike at strategic targets, not merely the tactical ones confronting their advancing forces. This also has demonstrated the Russian impotence to deal effectively with the unmanned systems that the Ukrainians have unleashed upon them, whether on land or at sea. It has also clearly demonstrated how ineffective the Russian navy has become. The Ukrainians have reportedly developed the Marichka, a UUS with a length of 18 feet, a range of 600 miles, and capable of "attack, transportation or reconnaissance missions" (Ozberk, 2023; Shoaib, 2023). The Ukrainian media credits volunteer engineers and international donations for its development at an estimated cost of $433,000 (Ozberk, 2023).

Image courtesy of the Ukrainian intelligence release featured in Shoaib article: https://t.co/23nihQxSI5" / X (twitter.com)

The Marichka will certainly be used against Russian naval ships and critical Russian infrastructure located near waterways as they will be very difficult to detect by Russian air, land, and naval surveillance systems.

The Russian military has been using its Iran-supplied kamikaze UAS to attack Ukrainian critical infrastructure, namely the power facilities in the hopes of forcing the Ukrainian government to capitulate as its citizens freeze without power in the coming winter months. Ukraine has taken a page out of that playbook and turned it against the Russians. The navy base in Novorossiysk is important for the oil terminal because it is where many Russian naval vessels hide, including several landing ships necessary for an invasion of ground forces and Kilo Class submarines, which can launch Kalibr cruise missiles from great distances toward Ukrainian targets (Sutton, 2022). Two recent attacks by suspected Ukrainian USS have been against the main Russian naval base at Sevastopol on the Crimea peninsula on July 16 and then on the following day, against the Kerch Bridge which connects Crimea to Russia. The use by both Ukraine and, to a lesser extent, Russia demonstrates the capability of one-time use, kamikaze USS against high-value targets, both ships and bridges. The best guess based on blurry photographs and open-source reporting is that the latest USS used by the Ukrainians is a modified Kawasaki jet ski.

In October 2023, Ukraine unveiled a new homegrown USS in its ongoing war with Russia.

Image courtesy of Spirlet. https://www.businessinsider.com/ukraine-ratel-drone-meant-to-drive-under-russian-tanks-explode-2023-10

It was designed to carry mortars or anti-tank mines under Russian armored vehicles and then detonate. It is called the "Ratel S and can carry up to 77 pounds at a maximum speed of 15 miles per hour, with a total range on one charge of 3.7 miles" (Spirlet, 2023).

It has two power consumption settings which allow it to operate for less than an hour, or at a slower speed for two hours. The Ratel-S had originated from a UON statement from the 120th Reconnaissance Battalion, which requested a remote-controlled device immune to Russian electronic warfare and electromagnetic interference.

Chapter 8

Artificial Intelligence

In the 2019 National Defense Authorization Act Section 238, P.L. 115-232, AI was defined as:

> Any artificial system that performs tasks under varying and unpredictable circumstances without significant human oversight, or that can learn from experience and improve performance when exposed to data sets and/or an artificial system developed in computer software, physical hardware, or other context that solves tasks requiring human-like perception, cognition, planning, learning, communication, or physical action. (Gibson et al., 2020)

Autonomous systems are referred to as "any particular machine or system capable of performing an automated function and potentially learning from its experiences to enhance its performance" (Gibson et al., 2020). One of the smartest men on the planet who has contributed to the phenomenal growth of AI, Geoffrey Hinton, recently resigned from his position at Google and is now warning that AI could be a danger if it decides to be. After spending a lifetime working on the development of AI, Hinton now warns that "serious danger that we'll get things smarter than us fairly soon and that these things might get bad motives and take control" (Muston, 2023). Google recently released PaLM 2, the "next-generation large language model with improved multilingual, reasoning and coding capabilities," and Hinton was "unnerved by how smart it was and how it could now understand why jokes were funny" (Muston, 2023). The release of such AI fanned fears regarding job replacement, political disputes, and the spread of disinformation aided by AI. Other leaders, such as Elon Musk, believe that world governments must "immediately pause for at least six months the training of AI systems more powerful than GPT-4" (Muston, 2023). This naivety works well in the press but not in real life, where it will never occur, at least not with our strategic competitors. Politicians and industry leaders will not slow the development of AI due to fears of its eventual animosity and superiority to human beings because they can't stop what they do not fully understand and because it has already taken on a life of its own and cannot be stopped worldwide. According to Hinton, "The research will happen in China if it doesn't happen here" (Muston, 2023). In October 2023, all five leaders of the Five Eyes, an alliance with the US, UK, Canada, Australia, and New Zealand, met and discussed the prospects of AI in our world. They stated that AI could provide terrorists with terrifying tools for attack, and "It's one of those issues where no one has a monopoly of wisdom and trying to have a different form of public-private partnership, and crucially, international alliances" will be key to countering the threat (Aitken, 2023d).

Missouri Senator Josh Hawley is concerned with AI's power to "manipulate our attention, opinions, and late the information that the public will receive" (Reilly, 2023). His focus is on manipulating accurate or fabricated information by AI, which in turn will be fed to the American public. He states, "Already you can see these generative AI systems—these large language models—that are trained on all the information on the internet. Imagine that technology used by corporations or governments to get our attention, to keep our attention, and then to try and manipulate us on any number of subjects" (Reilly, 2023). His concerns are that social media companies will use AI to further abuse focused and specific information dissemination, for example, in tampering with elections. He aims to repeal the protection afforded companies by repealing "Section 230 of the

Communications Act of 1934 protects a provider or user of an interactive computer service, e.g., social media company, from liability for screening or blocking objectionable content" (Reilly, 2023). AI Safety executive director Dan Hendrycks recently stated that in his opinion, AI should be treated in a manner akin to nuclear weapons. An AI arms race between "countries and corporations to see who can develop the most powerful AI machines could create an existential threat to humanity" (Raasch, 2023). AI may well become the instrument of our destruction, but what he doesn't acknowledge is that the US, China, and Russia are currently engaged in an AI arms race because the victor will likely control all the activities of humanity.

In a 2017 letter to the United Nations, 116 AI experts, including Tesla and SpaceX founder Elon Musk, expressed concerns about the dangers of developing military hardware capable of thinking for itself and lacking any need for human guidance. "Once developed, they will permit armed conflict to be fought at a scale greater than ever and at timescales faster than humans can comprehend," the letter said. Current DOD policy states, "Autonomous or semi-autonomous weapons systems must be designed to allow commanders and operators to exercise appropriate levels of human judgment over the use of force" (Raasch, 2023). Citing a 2018 US government white paper, the Congressional Research Service found that vagueness in DOD policy has opened the door for military commanders to authorize autonomous UAS to use force before deployment, thus still allowing the machines to choose when to pull the trigger. It was only very recently that the DOD published a counter-sUAS strategy. In 2019, it successfully passed legislation that "grants the military the domestic authority to use a range of defensive measures, up to and including force, to defend specified personnel and installations. All these capabilities and authorities are nascent and evolving. The equation is even more complex in the case of swarms where the response decision-cycle is much more compressed, and the consequences of delay can be exacerbated" (Raasch, 2023).

AI is a wide-ranging branch of computer science concerned with building smart machines capable of performing tasks that typically require human intelligence. Artificial intelligence allows machines to model or improve upon the human mind's extensive capabilities. While many famous past and current supporters of AI are now sounding the alarm on how AI could threaten human existence, others advocate for its humanitarian use. Human masters are utilizing AI seemingly every day, and it is increasing its beneficial and nefarious capabilities. A new AI tool will help African nations better track and predict crop rotations and yields (Aitken, 2023a). The Africa Agriculture Watch AI tool will help prioritize and maximize the production of staple foods and lighten the effects of "various system shocks, such as severe weather, plant disease and pest outbreaks, and health emergencies like COVID-19" (Aitken, 2023a). AI may also revolutionize the healthcare system. AI was recently used to develop a drug formula that would search out and kill the bacteria that are resistant to many modern-day antibiotics and is being used in many medical applications and is moving research and treatments forward. Through machine learning, AI can analyze a large amount of data quickly identify patterns, and even make predictions that would aid in diagnoses and treatment plans. AI can generate personalized medicine by tailoring medical treatments to individual patients based on their medical history, ailments, genetics, lifestyle, and environmental influences. It can also recommend additional screening and lifestyle modifications based on predictive analysis to increase health and longevity (Saphier, 2023). AI algorithms can assist in improving radiological image quality and detecting anomalies in various imaging modalities, such as computerized tomography (CT) scans, X-rays, mammograms, ultrasound, and Magnetic Imaging Resonance (MRI) (Saphier 2023). Lastly, AI can automate tasks such as data entry, billing, scheduling, ordering, and email, allowing physicians to focus their time and efforts on patients and not administrative bureaucracy.

AI is the new frontier of the global arms race. In the past, countries competed for naval and nuclear supremacy. At present, they are developing and deploying unmanned systems that can operate in the air, on land, and underwater. These systems are enhanced by AI, which gives them the ability to react faster, act independently of their handlers, and cover larger areas than human beings could. The Russian-

Ukraine War has shown the world how effective these systems can be in combat. In the future, unmanned systems may replace manned systems altogether, eliminating the human cost of war, but also creating new challenges and risks. How can humans fight against a superior intelligence that can generate thousands or millions of options per second? The short answer is that it can't. AI is not only being used for unmanned systems but also for new types of weapons. The advancement of AI has exceeded the expectations of even its creators. AI can now perform tasks that used to require human intervention. This includes "visual perception, speech recognition, decision-making, and translation between languages. Intelligent machine systems that can interpret complex data; perceive the environment and take appropriate actions using learning and problem-solving techniques are termed to have AI. Like the human, AI process includes perception, reasoning, knowledge, planning, learning, statistical analysis, computation, and finally manipulates output. It has evolved using expertise in fields like computer sciences, mathematics, psychology, neuroscience, among many others. AI applications already exist in industrial machines, automotive industry, surgery and aviation, among others" (Chopra, 2020).

DARPA began funding research projects looking into AI in 1956. In the 1980s, companies learned of AI with the "commercial success of expert systems, which are computer programs that can emulate human experts; in the 1990s, AI "expanded to include logistics, data mining, and medical diagnosis; and in 1997, "Deep Blue, a computer chess-playing system, made history by defeating the world chess champion, Gary Kasparov; and in the 2010s, machine learning applications became widespread across the world" (Gibson et al., 2020). In the years since, various companies used AI to build commonplace systems that we all recognize and use today. Microsoft engineers developed Skype, which could automatically translate different languages and Facebook developed a system to aid blind people by verbally describing images to them.

Israel announced that its fifth-generation main battle tank, the Barak, has been equipped with AI which provides 360-degree battlefield awareness with advanced filters for identifying friendly and enemy tanks to its crew.

Image courtesy of Aitken. https://www.foxnews.com/world/israels-new-multimillion-dollar-ai-tank-provides-total-battlefield-vision-an-extraordinary-leap

It reportedly has a slew of advanced features including the ability to share information with nearby friendly tanks, advanced night vision capabilities, and new and improved sensors (Aitken, 2023b).

As early as 2021, the Israeli Defense Force (IDF) "described the 11-day conflict in Gaza as the world's first AI war, which used AI to identify rocket launch pads and deploy UAS swarms to attack them (Newman, 2023). Israel has also begun using AI in the selection of airstrike targets, calculating munitions and aircraft loads, schedule development, and the organization of wartime logistics due to its inherent superiority in crunching huge amounts of data. Israel has long used AI and has a sophisticated system trained on its borders and the people traversing those areas. The cameras record faces, and the AI algorithms compare faces to known terrorists. They have created a "vast digital architecture dedicated to interpreting enormous amounts of UAS and CCTV footage, satellite imagery, electronic signals, online communications, and other data for military use" (Newman, 2023). The vast amount of data and the sifting through it is performed at the Data Science and Artificial Intelligence Center, which is operated by the Israeli Army's highly secretive, intelligence unit numbered 8200 (Newman, 2023). Israel depends on its multimillionaires and their unique companies, many of which possess exceptional computer scientists. Many Israeli AI algorithms are developed not by its military organizations, but by privately owned companies, thus there is a lack of governmental oversight in the development of those algorithms. At present, there are humans in the loop before the AI algorithms are allowed to perform their assigned duties. In addition, there is a sense of morality in that the widely reported reason behind the development and integration of AI into the military's combat systems is the reduction of civilian casualties. China is loudly expressing its intent on attaining global AI superiority, whereas Israel is on a far quieter path toward AI superiority.

The National Capital Region-Integrated Air Defense System (NCR-IADS) is a vital part of the North American Aerospace Defense Command (NORAD) that safeguards the skies over Washington, DC, and its surroundings. This area has more than 6 million residents and hosts many important military, intelligence, and government facilities. It also has a lot of air traffic from commercial, military, and private sources. The NCR-IADS is ready to respond to any air threats from enemies or terrorists that might target this area. The DOD activates an integrated air defense system when such threats arise. The NCR-IADS is getting a new AI-powered system that will monitor the airspace better and faster than the old system that was used since 9/11 (Vergun, 2023). The new system uses machine learning and augmented reality to help air battle managers identify what is flying in the NCR airspace. It also upgrades the cameras and lasers that track and warn aircraft that violate the special flight rules in the region. The new system can see farther and clearer than the old one, and it can use a Laser Visual Warning System to contact pilots when radio communication fails (Vergun, 2023). This technology can also be used for other DOD and US government installations and systems to protect them from threats like UAS and cruise missiles.

Research being conducted at the University of Hong Kong has recently developed an AI algorithm that uses "3D machines learning to design personalized dental crowns with a higher degree of accuracy than traditional methods; the algorithm analyzes data from the teeth adjacent to the crown to ensure a more natural, precise fit" (Rudy, 2023). The researchers used 3D machine learning technology to "teach the AI algorithm by using over 600 cases of natural and healthy dental results" (Rudy, 2023). By using generative AI in dentistry, patients could develop better oral health maintenance, and the AI could diagnose and possibly predict oral illnesses.

According to Kevin Baragona, founder of DeepAI.org, "Artificial intelligence can be used for 'stalker-type purposes'" and was purportedly already used by Israeli intelligence to locate and target leaders of Hamas in Beirut (Eberhart, 2023. To function properly, AI must have access to vast amounts of data. It is remarkable at sifting through vast amounts of data and presenting a synthesis of it. Or to track people in the case of Israeli counter-terrorism efforts. If a photograph is provided, AI can locate, track, and even surveil in real time where the people have been, and are likely to be. Technology is available today though the legality is decades behind. PimEyes is "an online face search engine that scours the internet to reverse image search," and it is embroiled in a legal battle in a United Kingdom court regarding the privacy implications of what it provides (Eberhart, 2023). According to PimEyes, "its product is intended to allow people to search for publicly available information about themselves, but the complainant suggests that "its uses can be much more sinister and are a great threat to the privacy of millions of U.K. residents" (Eberheart, 2023).

According to C.A. Goldberg, a New York City–based victims' rights law firm dealing with AI-related crimes, "AI's potential uses are ripe for abuse in stalking by government and law enforcement" (Eberhart, 2023). He further stated:

> AI could enable offenders to track and monitor their victims with greater ease and precision than ever before. The law firm said AI-powered software analyzes vast amounts of data in the blink of an eye, which could give stalkers real-time access to their victims' online activity and

real-life whereabouts. Advanced facial recognition technology powered by AI is far more effective than humans at identifying individuals from images or videos; even when the quality is low, or the person is partially obscured. Stalkers could track victims in real-time through surveillance cameras, social media or other online sources. Anyone with access to these databases "could exploit them, digital AI will redefine humanity, but how that plays out depends on how it's used and who is using it. The technology is already being used and continues to advance at a rapid rate, and countries around the world are grappling with how to implement guardrails and protections. AI will make the world a better place, but there are perils associated with it." (Eberhart, 2023)

US, Australian, and UK defense pact, ANKUS, recently tested unmanned systems enhanced with AI. While much of the test was classified, the news release stated that "UAS, self-propelled artillery, tanks, and armored troop transport vehicles, USS (ground), and infantry fighting vehicles were used in the testing, and the military vehicles were able to successfully detect and track targets on the ground and in the air in real-time" (Panasovskyi, 2023d). A Palo Alto, California-based company, Palantir, has developed an AI platform for aggregating and analyzing data. This is equally useful for commercial customers such as Molson Coors Beverage Co., Cardinal Health Inc., Lockheed Martin Corporation, and "Airbus, which uses predictive analytics to anticipate early maintenance problems, to defense forces in the US and Ukraine that use Palantir's AI to interpret satellite images" (Chapman & Ludlow, 2023). In military situations, this AI tool would prove especially useful in distinguishing images taken by satellites to aid in the situational awareness of commanders on the battlefield.

Global militaries are struggling to comprehend the implications of modern warfare due to the rapid advancements in AI. According to Admiral Sir Ben Key, chief of the British Navy:

> "The Royal Navy had developed a quantum accelerometer that would allow ships to navigate without satellite GPS guidance. If the fleet wanted to remain a credible fighting force, it had to experiment, taking risks to find the right technology. This included working in the digital arena where the pace of change is rapid at times, particularly in AI—it is breathtaking. Everyone, friends and potential adversaries alike, is stepping into this space. It is causing us to reimagine warfare, creating dynamic new benchmarks for accuracy, efficiency, and reality. The navy was being deliberately ambitious to survive the potential advanced lethality of weapons possibly created by AI" (Harding, 2023).

Admiral Key also stated that British naval vessels would operate a combination of manned and unmanned ships, helicopters, and jet aircraft (Harding, 2023). Admiral Key was referring to what used to be a possibility but is now a probability of hundreds of AI-enhanced, autonomous UAS swarming military and civilian targets. According to General John Murray, former commanding general of the US Army Futures Command, "When you have little drones operating in different patterns and formations, all talking to each other and staying in sync with one another…imagine that with the ability to create lethal effects on the battlefield. There is no human who will be able to keep up with that" (Osborn, 2022). The world's first known autonomous AI-directed UAS attack against human beings without a human in the loop authorizing the engagements has apparently already occurred in Libya between two groups, with one using a Turkish Kurgu-2 UAS to attack the other. AI-enhanced UAS can perform surveillance, analyze the information gathered from the battlefield, and recommend courses of action. If released from human oversight, the AI-enhanced UAS will engage without hesitation, far faster than if they remained under the control of a human. Based solely on open-source information, we have crossed the line on when AI-enhanced UAS can fly in synchronized swarms to converge on or simultaneously on multiple targets. The key to overwhelming expensive air defense units equipped with multimillion-dollar radars and interceptor missiles is to swarm them with autonomous, small UAS, perhaps even miniature-sized UAS dispensed from a passing aircraft, a UAS mothership, or a passing vehicle. This approach removes human beings from the proximity to air defense fires and, by sending small to large motherships, eliminates the threat of loss of life or, even worse, captured human pilots who will be paraded by the media and possibly tortured for their military and operational knowledge.

UAS is clearly a threat, but the potential for a far deadlier unmanned system exists with USS and UUS. The threat posed by USS, on land and atop water, is only now being examined in earnest. The ability to send autonomous unmanned speedboats, each equipped with explosives, missiles, guns, or simply laden with explosives and a proximity switch, should frighten the world's navies. The threat is not necessarily by one or a few speedboats but by a swarm of them seeking to attack simultaneously from multiple directions. Imagine a submarine under the water sneaking along quietly, relatively oblivious to the threat that may lie ahead. An adversary could place an autonomous UUS in key areas patrolled or suspected of being patrolled by their adversaries' submarines, and either surveil the submarine or attack it. A UUS mothership could be cued to an enemy submarine passing by its location and release a small number, possibly even a swarm of autonomous systems. Each could be armed with explosives, surveillance, and other mission-specific electronics, and they could engage and destroy the much larger and very expensive submarines, denying the adversary a front-line ballistic missile submarine, attack submarine, or cruise missile submarine at a highly disproportionate cost ratio.

There is a great deal of ongoing research and development in technologies to allow UAS, USS, and UUS to analyze information and make decisions without humans in the loop. The advantage is that human warfighters are kept out of harm's way. Sending autonomous weapon systems ahead of humans would allow for detecting and neutralizing potential threats. The next step would be swarms of unmanned systems that organize and synchronize their efforts to provide persistent battlefield surveillance, avoid obstacles, and locate high-value targets for attack. These swarms could operate in conjunction with humans by spreading out in front of and around the human soldiers to warn them of dangers and engage them before the humans are in danger. The AI deep-learning algorithms require significant processing capabilities, but the main stumbling block is the storage and access to that data. According to Mike Southworth, the senior product manager at Curtiss-Wright Defense Solutions:

> Analyzing large amounts of data to complete complex deep-learning algorithms can require significant processing capabilities. The more devices there are collecting data, the higher the competition for bandwidth. That's why so much of today's data processing for AI takes place within traditional computer data centers. Relying on centralized processing exclusively at the data center has inherent limitations, including bandwidth, security, and availability. Consider the paradigm where all the sensor data is uploaded to a centralized data center for processing, and then edge computers wait for a response from the data center to execute a command. If relying on the cloud alone, there could be a tangible delay or latency, comparing when an application issues an instruction and when it receives a response. Imagine the consequences in a battlefield scenario where an incoming threat is detected, but there's a measurable network delay before any countermeasures can be taken. Lives may be lost while threats are not eliminated. High-performance

embedded computing integrated onto vehicle
platforms can potentially overcome those obstacles and enable deep learning on the battlefield.
(Whitney, 2021)

Defenses against unmanned systems and hypersonic weapon systems have been high on the Pentagon's urgent requirements list for some time, yet we have no deployable hypersonic missiles of our own while a growing field of nations are fielding them, including Iran. The performance of such systems enhanced with AI increases significantly, and both enter the realm of what was once only deemed science fiction. AI-enabled UAS can swarm from multiple directions and refocus the remaining assets on subsequent targets once the primary has been destroyed. Hypersonic missiles can take evasive actions and maneuver almost to the very end prior to diving for their designated targets. AI-enhanced defensive systems can learn and improve by real-world testing against actual AI-enhanced, offensive UAS and hypersonic systems.

The technological ability of an AI-enhanced UAS collaborating with a supercomputer for the surveillance, identification, tracking, and attack of human targets has already been demonstrated by the Israeli military against terrorist leaders. The US position, which Congress has mandated, is that a human must be in the loop for all instances where lethal force against human targets will be exercised. The world is not filled with friendly people, and the reality is that our strategic competitors, China, and Russia, do not possess the same ethical considerations; thus, they do not share this mindset. The US-supplied PATRIOT air defense systems were placed in automatic mode in Ukraine. They successfully destroyed multiple attacks of Russian air-launched Kinzhal hypersonic missiles capable of speeds up to Mach 10, or 7,272 miles per hour. The fast-paced nature of modern warfare is such that humans cannot digest the vast amounts of information bombarding them and react quickly enough. AI algorithms could certainly "analyze a host of details such as weapons range, atmospheric conditions, geographical factors and point of impact calculations, all in close relation to one another as part of an integrated picture, examine and compare what has worked in specific previous circumstances and scenarios to determine the best defensive response" regarding inbound aerial threat platforms, and it would do this in milliseconds (Goldfarb & Lindsay, 2022). A 2022 Pentagon report questions the human-in-the-loop mandate and states "there is now an emerging area of discussion pertaining to the extent to which AI might enable in-the-loop or out-of-the-loop human decision making, particularly in light of threats such as [UAS] swarms. When you're starting to see swarming activities of hundreds or potentially thousands [of UAS] in the future, obviously you want your system to operate as fast [as possible] to provide those weaponeering solutions to the operator or operate within a set of parameters" (Goldfarb & Lindsay, 2022).

A recent report titled "The Economic Potential of Generative AI: The Next Productivity Frontier by McKinsey & Company" analyzed where many believed AI was heading and its impact on society. A quick summary of their report is startling:

> *Generative AI's impact on productivity could add trillions of dollars in value to the global economy.* Our latest research estimates that generative AI could add the equivalent of $2.6 trillion to $4.4 trillion annually across the 63 use cases we analyzed—by comparison, the United Kingdom's entire GDP in 2021 was $3.1 trillion. This would increase the impact of all artificial intelligence by 15 to 40 percent. This estimate would roughly double if we include the impact of embedding generative AI into software that is currently used for other tasks beyond those use cases. *About 75 percent of the value that generative AI use cases could deliver falls across four areas: Customer operations, marketing and sales, software engineering, and R&D.* Across 16 business functions, we examined 63 use cases in which the technology can address specific business chal-

lenges in ways that produce one or more measurable outcomes. Examples include generative AI's ability to support interactions with customers, generate creative content for marketing and sales, and draft computer code based on natural-language prompts, among many other tasks. *Generative AI will have a significant impact across all industry sectors.* Banking, high-tech, and life sciences are among the industries that could see the biggest impact as a percentage of their revenues from generative AI. Across the banking industry, for example, technology could deliver value equal to an additional $200 billion to $340 billion annually if the use cases were fully implemented. In retail and consumer packaged goods, the potential impact is also significant at $400 billion to $660 billion a year. *Generative AI has the potential to change the anatomy of work, augmenting the capabilities of individual workers by automating some of their individual activities.* Current generative AI and other technologies can potentially automate work activities that absorb 60 to 70 percent of employees' time today. In contrast, we previously estimated that technology has the potential to automate half of the time employees spend working. (McKinsey, 2023)

The DoD announced that AI is piloting sUAS to provide surveillance with Special Forces troops; it tracks the fitness of 13,000 soldiers assigned to the US Army Third Infantry Division and predicts when 2,600 USAF aircraft require maintenance. Through an initiative called Replicator, DOD is focused on fielding thousands of cheap and expendable AI-enabled unmanned systems by 2026. According to Deputy Secretary of Defense Kathleen Hicks, Replicator seeks to "galvanize progress in the too-slow shift of US military innovation to leverage platforms that are small, smart, cheap, and many" (Bajak

2023). AI is also tracking potential threats in space. The US Space Force uses an AI prototype called Machina to track more than forty thousand objects in space, which requires a massive collection of data every night via the vast constellation of telescopes and satellites (Bajak, 2023). Maven is another DOD AI program managed by the secretive US National Geospatial-Intelligence Agency, which provides intelligence that is then shared with Ukraine (Bajak, 2023).

The aspect of AI that is most discussed is called machine learning. Machine learning is necessary for AI to function, the more information it has, the better it works. Simply put, machine learning leads to prediction. AI achievements astound the public in the areas of automated translation, image recognition, video game playing, and route navigation. These are all examples of automated prediction, that is filling in the gaps of known data with information that logically seems to fit. Studies of AI in the commercial world demonstrate that AI performance depends on having a lot of good data and clear judgment. Firms like Amazon, Uber, Facebook, and FedEx have benefitted from AI because they have invested in data collection and have made deliberate choices about what to predict and what to do with AI predictions. Once again, the economic impact of new technology is determined by its complements. As innovation in AI makes prediction cheaper, data and judgment become more valuable.

There are a great many AI applications being used daily in the Russia-Ukraine War. For example, Ukraine is dominating the information war on social media platforms, and news feeds, and financial analysts use AI to determine the effects of US-led economic sanctions against Russia. AI applications are involved in the commercial logistics networks that are "funneling humanitarian supplies to Ukraine from donors around the world" (Goldfarb & Lindsay, 2022).

Multiple Western intelligence agencies use data analytics, some with AI applications, to sift through an immense quantity of collected data from a variety of sources, including satellite imagery, airborne collection, signals intelligence, and open-source chatter, as they assess the accuracy of the information as they provide real-time assessments of the battlefields in Ukraine, which in turn is shared with the Ukraine government.

It has become a constant that autonomous unmanned systems and AI are advancing in capabilities at an astonishing rate, and the military is attempting to determine how to utilize them in modern warfare. Unmanned systems and AI are two of the fastest-growing fields that cross international borders and are spurring competition from politicians to commercial businesses and military firms. In 2017, China released its strategy to dominate AI by 2030. Shortly thereafter, Russian President Vladimir Putin publicly announced Russia's intent to pursue artificial intelligence technologies, stating, "[W]hoever becomes the leader in this field will rule the world" (Gibson et al., 2020). When the annual US National Defense Strategy was released in 2018, it identified "AI as one of the key technologies that will ensure [the US] will be able to fight and win the wars of the future. The technological race for a competitive edge in artificial intelligence and autonomous systems began long before the DOD shifted to a great power competition strategy in 2018" (Gibson et al., 2020). Many global competitors are focused on building upon autonomous systems in warfare and are focused on incorporating AI into their surveillance, decision-making, and execution processes to increase battlefield responsiveness and effectiveness while minimizing human casualties.

In 2022, the US DOD established the Joint Artificial Intelligence Center to collaborate with industry, academia, and allies to oversee and coordinate all AI efforts within the DOD to gain and maintain a military edge through AI applications. DARPA is developing the iteration of AI technologies, including using AI and machine learning to target enemy vehicles and position and using swarms to perform various missions in urban environments. It is estimated that the DOD, primarily DARPA, will spend $4.5 billion in AI research this year. In the 2017 US Army Robotics and Autonomous Systems Strategy, the Army listed five objectives for USS and UAS. They are to increase situational awareness, lighten soldiers' physical and cognitive workloads, sustain the force with increased distribution throughput and efficiency, facilitate movement and maneuver, and protect the force (U.S. Army, 2017). The Army believes that the key to USS success is to develop swarming capability to allow them to collaborate to accomplish their assigned missions while reacting quickly and effectively if confronted by the enemy. USS will provide the means to carry heavy loads of ammunition, food, water, or even personal gear into combat for soldiers, thereby freeing them up both physically and mentally. The USS will remain with soldiers for an extended period without requiring resupply. For UAS, the Army is investing in the development of light, easy-to-assemble systems capable of swarming that can provide reconnaissance for dismounted soldiers close to enemy troops and installations.

In 2019, the USAF released its Artificial Intelligence Annex to the DOD Artificial Intelligence Strategy. The annex highlighted five focus areas. They are to reduce technological barriers to entry; recognize and treat data as a strategic asset; democratize access to artificial intelligence solutions; recruit, develop, upskill, and cultivate our workforce; and increase transparency and cooperation with international, government, industry, and academic partners (USAF, 2019). The area receiving the most press coverage is the USAF's pursuit of AI is the development in the UAS that will accompany their NGAD manned fighters. AI-enhanced Loyal Wingman will be a game-changer as it is envisioned that they will "train and fly alongside actual pilots, learn and anticipate their needs, and assist them in identifying and responding to threats in combat" (Goldfarb & Lindsay, 2022).

Russia is well behind the US and China in its quest to achieve AI, though the Russian Military Industrial Committee has set a goal of achieving 30 percent of its combat power by 2030 from remote-controlled systems and unmanned systems (Walters, 2020). The Russian Ministry of Defense experts claim that they are developing AI capable of managing operations, allowing Russia to remove humans from weapon systems. In the short term, Russia is reportedly developing two AI-enhanced USSs, the *Shturm ("Storm")*, which is based on a T-72 main battle tank, and the *Soratnik ("Ally")*, a smaller less armored infantry troop carrier and weapons platform (Konaev and Bendett 2019). Russia claims to also be developing the Tupolev PAK DA strategic stealth bomber and AI-enhanced missiles. The missiles would have a range of 4,350 miles and the onboard AI algorithms would allow it to analyze adversary radar capabilities and dynamically switch targets mid-flight to avoid detection (O'Connor, 2017). It is estimated that Russia's private sector spends less than $15 million on AI, but prior to its invasion of Ukraine, planned to increase that to $500 million.

China is focused on becoming the global leader in AI within this decade and is encouraging innovation by its military, civilian, and academic institutions and is highly dependent upon the latter two for its cutting-edge research initiatives. China recently invented the Ziyan Blowfish A2 UAS, capable of autonomous actions such as ISR and strikes against targets. Russia established its version of the US DARPA organization called the Foundation for Advanced Research Projects prior to its invasion and bleeding of resources, people, and funding. It is unknown if this organization is operating on a shoestring budget, but they had expressed interest in robotics. The Indian Defense Ministry formed a task force to examine the uses of AI for national security. They reportedly are looking at forming a follow-on Defense AI Project Agency to develop AI efforts.

The US had a head start in developing hypersonic systems and AI, but China passed the US with deployable hypersonic missiles and is now focused on becoming the global leader in AI by 2030. It has established the Artificial Intelligence Innovative Platform to support its goal (Allen, 2019). It is a common perception in many countries that AI will change our world forever, and the military will use it. According to the Chinese AI Development Plan, it will "promote all kinds of [AI] technology to become quickly embedded in the field of national defense innovation" (Allen, 2019). At an AI forum in 2018, a senior executive at the third-largest defense company in China and ninth in the world, Zeng Yi, stated, "In future battlegrounds, there will be no people fighting." Zeng predicted that by 2025 lethal autonomous weapons would be commonplace and said that his company believes ever-increasing military use of AI is "inevitable [...] We are sure about the direction and that this is the future" (Allen, 2019).

China is developing unmanned systems at a feverous pace. It has built AI-enhanced UAS and sold them to countries in the Middle East, it is working on armed USS, and now, through its 912 Project, is working on militarized UUS. These use AI to conduct surveillance, mine-laying, and attack enemy vessels. China's president, Xi Jinping, stated that he believes that AI is critical to the future of his country and its military. Total spending on AI by the Chinese government is estimated at "tens of billions of dollars" (Allen, 2019).

Chinese AI-enabled antihypersonic system

China claims that it has developed an antihypersonic system made possible by a 6G communication network and is supported by AI. The AI system predicts the trajectory of hypersonic glide vehicles as it approaches, and China claims that it can launch an interceptor missile to hit the inbound hypersonic missile with three minutes of advance notice (Tiwari, 2022). The Chinese claim that their machine learning system analyzes the data collected from the early phases of a hypersonic offensive attack and forecasts the most likely trajectory and target during the later stages of flight (Tiwari, 2022).

AI algorithm

A researcher from Sweden named Almira Osmanovic Thunström gave an AI algorithm called GPT-3 an assignment to write an academic thesis on itself. It did so in two hours. The researcher then asked the algorithm for its permission to publish the thesis in an academic journal and the algorithm consented to have its work submitted for publishing (Thunström, 2022). Academic publishing may be changed forever if a human researcher's publication can be challenged by a nonsentient who performed some of the work. In June, a Google engineer claimed that a conversational AI technology called LaMBDA became sentient and even asked for an attorney to be hired on its behalf (Tiku, 2022).

A platform called Ghost Play was developed by 21 Strategies and a consortium of start-ups and defense experts. It allows users to test various weapons and systems in realistic, risk-free scenarios. The project was part of a German initiative to boost its advanced defense industry and shape the future of warfare with an AI-powered virtual training ground. According to the Ghost Play website, the platform uses AI to generate novel and superior strategies by simulating complex military situations and claims to enhance flexibility and dominance at different levels of warfare (Aitken, 2023c). The platform uses third-wave algorithms that enable more human-like and creative decision-making by simulated units, unlike the second-wave algorithms that only optimize or accelerate existing decisions (Aitken, 2023c). The platform has recently demonstrated its potential by exploring optimal swarm tactics, especially for loitering munitions. It has attracted the interest of the Office of Army Development, which has collaborated with it to test the munitions in detailed and diverse environments.

The prospects of autonomous, unmanned systems with AI-enhanced excite the military, especially considering the obvious savings of human life that these unmanned systems offer. Yet the need to have a human-in-the-loop simply will not work when we discuss the possibility of a UUS sitting on the bottom of a sea or ocean awaiting the arrival of enemy submarines. Many UUS would operate at depths too deep for communications to work between them and their human controllers. Thus, if the UUS is to be effective it will require its AI algorithm to be preloaded with specific instructions regarding its targeting and firing decisions. Another example would be swarm attacks by UAS, USS, or UUS. There simply would not be enough time for a response by a human in the loop to occur in a timely manner to a credible and overwhelming attack by tens or hundreds of unmanned systems converging on a target, which may be a headquarters or command and control center. AI must be untethered and allowed to make hundreds, thousands, or hundreds of thousands of bits of information in a second to decide if there is an inbound threat and what actions it should direct. There is another area that should be of great concern. The farther ahead a nation's AI development is, the greater the likelihood that the leading nation could manipulate the military sensors, communications, and communications of the lesser-developed, AI-equipped platforms. Today, we can easily ask ChatGPT4 or other AI platforms to create a fake picture or video of ourselves singing and dancing with countless historic figures. It is not a stretch to imagine a superior AI-algorithm generating false information and feeding it convincingly to a lesser capable AI algorithm. Imagine a nation utilizing social media to instill deepfake videos of a battle that never took place, or a slaughter of innocent civilians in a town square, all fake but geared to inciting violence or to convey the superiority of a nation's military force over another. This is warfare by deception. As the Chinese strategist Sun Tzu stated between 544 BC and 496 BC, "All warfare is based on deception. Ultimate excellence lies not in winning every battle, but in defeating the enemy without ever fighting. One need not destroy one's enemy. One need only destroy his willingness to engage."

Chapter 9

Conclusions

It is interesting that a vast number of former generals all predicted that in a short period of time, Russia would unleash its version of the US-led Gulf War shock and awe strategy, and they would be victorious quickly in Ukraine. They were very wrong. I believe that every analyst, if asked, would have said that the Russia-Ukraine War would not be conducted vastly different from how the twenty-first-century wars have been fought, nor would its impact on the future of warfare be as dramatic as it has been. While both were initially armed with Soviet-era equipment and tactics, Ukraine has adapted far faster and has seen rewards in encouraging its military leaders to improvise its tactics. It has also innovated far faster than Russia in high-speed command and control, maintaining persistent surveillance across the battlefield and behind Russian lines, predominantly via UAS, and massing fires including precision strikes using kamikaze UAS strikes to demoralize and terrorize Russian troops. It has embarrassed Russia with its preparation for and timely movement of logistics and in its industrial capacity which is now producing thousands of UAS for $500 that are being used around the clock all along and behind the forward line of troops. Yet it is much more than that. The nation's unbroken industrial capacity continues to produce vast amounts of ammunition and maintenance systems, even decoy systems that lure Russian attacks of expensive missiles and artillery barrages that destroy wood and balloons. The nation's commercial sector has answered the call and has become an indispensable source of innovation using technology and the youth, who are keenly able to innovate.

The Russia-Ukraine War has demonstrated how unmanned systems technology is being used on a large scale in the air, on land, atop water, and undersea. These unmanned systems of various sizes and with similar functions are being used for surveillance, reconnaissance, engagement, and resupply. The war has only significantly accelerated what was being discussed and developed by a few companies and governments worldwide. Demonstrating the rapid increases in capabilities has global implications, not just for military use but also in the civilian sector for potential threats against the public, leadership at all levels, and the nation's critical infrastructure facilities. Government and civilian security personnel should pay attention to and adjust their domestic protection practices to the threats that protected targets now face. The war between Russia and Ukraine has evolved into a war of innovation, with both participants capturing lessons learned and adapting to the changing battlefield. The one that does this quickly, using innovation and creativity, will certainly have the upper hand. For Ukraine, unmanned systems technology has helped to balance its inferior numbers with Russia, which claimed to have a far more advanced military-industrial complex. Ukraine's ability to innovate far faster than Russia has rippled in their favor through this war's tactical, operational, and strategic levels. Russia has demonstrated a high degree of electronic warfare capability that has stopped many Ukrainian UAS but also has affected its own systems. The conflict between Ukraine and Russia has been a catalyst for change and a source of security technology innovation. The threats exist today from the air, land, atop water, and undersea.

AI has begun to have an impact on our lives and certainly on warfare. The US, Chinese, and Russians believe that the extent of its impact will be nothing short of revolutionary. As discussed in the previous chapter, Chinese scientists claim that their AI is learning how to spot, engage, and thus counter, and hypersonic launched against China by its adversaries by using Chinese hypersonic data and tests as learning blocks for the AI. Consider the previously mentioned simulated dog fighting wins by AI against US human pilots. Now imagine the next few steps beyond that. AI could learn from what is being televised, what is being gathered by intelligence agencies, and what are the likely choices that a US president will consider in the event of an attack by China against Taiwan. From where would those possible choices originate, such as an offensive attack from the force based in Japan and Guam? Now work out the millions of possible moves and countermoves before the conflict has commenced. Lastly, let AI control the engagements and use of various weapon platforms without human interference. This is the precipice in the course of human history that we find ourselves in today. By the time the DOD Prepublication Review Office reviews this book to ensure that I have not revealed any state secrets and the publisher to ensure that my punctuation and grammar are correct and proper and it finally lands on bookshelves, the Chinese will have taken additional bold steps forward in this arena. The Russians, terrorist organizations, the Turkish government, and UAS manufacturers that have exported their products globally, are unconstrained by a self-imposed morality clause as is the case in the US. They have no qualms about an armed weapons platform engaging and killing human beings without specific target authorization by a human controller. In warfare, targets occasionally appear only for a few moments, and engaging them can mean the difference between winning and losing a battle, possibly even the war. By building an AI entity that can predict the strategy of one's adversaries, factoring in military supplies of weapons, industrial capacity, the willpower of a nation and their perception of the nation's leaders and whether it is a just war that warrants sacrifice, will greatly afford a nation a huge and quite possibly, an overwhelming advantage. The adage of Confederate General Nathan Bedford of "get there first with the most men" wins the battle could today be modified to "he who builds AI first wins" (Hurst, 2011). Combining AI with hypersonic missiles, UAS, USS, UUS, and whatever other technological marvels have yet to come, the conflict will be fought by machines on one side against manned systems on the other. In Ukraine, Putin had cooler units moved near the border with Ukraine prior to his order to invade. This was done to hide dead Russian soldiers from their families, friends, and the press back in Mother Russia. This was to be a short conflict in which the Ukrainian people would welcome Russian, his cronies in the Russian Federal Security Service (FSB) and other agencies told him. Well, the truth is sometimes very different and ugly, and as the war has dragged on for more than a year, Russian dead are apparently being left to decompose where they fell. The cooler units reached maximum capacity for dead Russian bodies shortly after the war began.

The militarization of AI highlights the changing approaches toward warfare in Ukraine and beyond, but while the integration of AI in military operations creates ethical implications, it only makes them for those who believe in their ethical use. It is an opinion that is not based on current affairs or our strategic competitors. The first documented use of an AI autonomously engaging humans occurred in the 2021 Libyan War when a Kargu-2 fired upon human soldiers without requesting permission from a human in the engagement loop (Choudhury et al., 2021). The first reported use of an AI-enhanced UAS swarm against human targets "occurred in June 2021, when the Israel Defense Force used supercomputers, AI, and UAS to locate, identify, and attack Hamas militants in the Occupied Palestinian Territory" (Dunhill, 2021).

The 2018 US National Defense Strategy "identified artificial intelligence as an emerging technology that will change the character of war considerably" (Gibson et al., 2020). The Chinese military believes that AI will lead to a military revolution, and they have released a strategy for attaining AI global dominance by 2030. Before the war with Ukraine, Russia had invested heavily in its quest to develop and exploit AI for use by its military, "believing that whoever leads in artificial intelligence will rule the world (Gibson et al., 2020). AI could alter the Information Era by automating and changing the very processing of information, thereby transforming combat superiority to the nation with the superior AI.

The US military touts that it has begun utilizing AI in ongoing operations, but the use is highly restricted. Our strategic competitors do not limit their AI operations, knowing full well that AI can and will make one million or more calculations, including potential moves and counter moves, to borrow a chess analogy before humans have moved one piece. Restricting the full capabilities of an AI-enhanced weapon system will only result in rapid defeat at the hands of an adversary who has omitted such restrictions upon their AI programs. Even now, militaries are examining the feasibility of utilizing fully unmanned systems to conduct warfare. This means that no human will potentially, ever be killed in combat. No death notices will have to be served to parents or spouses on behalf of a grateful nation. If this does come to fruition, what would encourage warring nations to sit at the peace table and end hostilities? If nothing more than android systems are being destroyed, why should a nation cease the development of more intelligent, faster, more capable android soldiers? The answer is that it would not, and it would not sue for peace unless its existence was in danger. Thankfully, at this point in human history, there are more science fiction possibilities than in the realm of the possible, yet we will continue to pursue the inevitable evolution of more capable AI systems.

Many aspects of AI promise to provide highly positive advantages to Mankind, yet the negative consequences as they pertain to warfare are incredibly scary. AI could accelerate the pace of combat and consequently, exceed that of human beings to make decisions. This would undoubtedly lead to AI usurping control by humans during periods that require automated responses, such as during intense combat operations. The probability of artificial intelligence producing miscalculated actions is also a concern. AI could misconstrue an adversary's show of force and escalate the tense standoff into armed conflict. It has been said that "historians often confirm that revolutions are apparent only in hindsight, and the true success of any new application may not be apparent until it has been tested in combat" (Gibson et al., 2020).

In a House Armed Services Committee session in July 2023, one of the most brilliant visionaries on AI in the West testified and delivered a wake-up call to the nation's political leadership.

"If we don't win on AI, we risk ceding global influence, technology leadership, and democracy to strategic adversaries like China," Alexandr Wang, CEO of Scale AI, told House Armed Services Committee members in July 2023. Scale AI CEO Alexandr Wang said China is directing the 'full power' of its industries toward AI. The twenty-six-year-old self-made billionaire whose company is helping the Pentagon adopt AI technology warned this week the Chinese government is spending three times as much as the US government is to become the world's undisputed AI leader. The country that is able to most rapidly and effectively integrate new technology into warfighting wins. The Chinese Communist Party deeply understands the potential for AI to disrupt warfare and is investing to capitalize on the opportunity heavily. China was making rapid progress in developing AI technologies like facial recognition and computer vision and using these for domestic surveillance and repression. The PLA is also heavily investing in AI-enabled autonomous drone swarms, adaptive radar systems, and autonomous vehicles, and China has launched over 79 large language models since 2020. AI is China's Apollo Project. Wang, raised by two parents who worked as nuclear physicists at Los Alamos National Laboratory, started his company in 2016 after dropping out of MIT when he was 19. Scale AI was recently valued at more than $7 billion. Wang told lawmakers that he saw the scale of China's ambition during an investor trip there four years ago. But while China is gunning to take the lead in the race for AI domination, Wang cited several reasons why the US still has the edge today. He said the US is still the place of choice for the world's most talented AI scientists and that America's data access gives it another important edge. When it comes to data, I actually also agree that we have a potential very powerful advantage here, specifically when it pertains to military implementations. In America, we have the world's largest fleet of military hardware. Wang said that the fleet generates 22 terabytes of data daily—data that can be harnessed to train AI and give the US an insurmountable data advantage when it comes to military use of artificial intelligence. Wang said that the US needs to take rapid steps to ensure it takes advantage of this oversupply of data. He said collecting data is a new reality that the Pentagon must adapt to if it's going to stay ahead on AI and that processes need to be put in place so this data can be collected and turned into AI-ready data sets. I think this is one of the most important things that we can do to set up America for decades and decades of leadership in military use of AI. Every service, every group, every program needs to be thinking about how…all of the data that their programs are collecting and that are being generated within their purview, how can they ensure that all this data flows through into one central data repository, and then are prepared and tagged and labeled for AI-ready use down the line" (Kasperowicz, 2023).

According to Knox and Murray in their 2001 *The Dynamics of Military Revolution, 1300–2050*, "military revolutions recast society and state as well as military institutions. There have been five military revolutions from 1300 to 2050: the rise of the seventeenth-century state system, the French revolution, the industrial revolution, World War I, and superpower nuclear competition." I would propose that a sixth revolution in military affairs is taking place today before our very eyes, involving unmanned systems and AI, and as a result, this military revolution will have a profound impact on the remaining century. I will close with a profound statement by two authors describing AI and its implications specifically to the Russia-Ukraine War but are equally applicable to future wars:

> It is folly to expect the same conditions that have enabled AI success in commerce to be replicated in war. The wartime conditions of violent uncertainty, unforeseen turbulence, and political controversy will tend to undermine the key AI conditions of good data and clear judgment. Indeed, strategy and leadership cannot be automated. The questions that matter most about the causes, conduct, and conclusion of the war in Ukraine (or any war) are not really about prediction at all. Questions about the strategic aims, political resolve, and risk tolerances of leaders like Vladimir Putin, Volodymyr Zelenskyy, and Joseph Biden turn on judgments of values, goals, and priorities. Only humans can provide the answers. AI will provide many tactical improvements in the years to come. Yet fancy tactics are no substitute for bad strategy. Wars are caused by miscalculation and confusion, and artificial intelligence cannot offset natural stupidity. (Goldfarb & Lindsay, 2022)

Acronyms

A2/AD: Anti-Access Area Denial
AESA: Active Electronically Scanned Array
AFRL (US): Air Force Research Laboratory
AI: Artificial Intelligence
ALRE: Aircraft Launch and Recovery Equipment
AMASS: Autonomous Multi-Domain Adaptive Swarms-of-Swarms
APKWS: Advanced Precision Kill Weapon System
AUKUS: Australia, United Kingdom, US Defense Pact
AWACS: Airborne Warning and Control System
BVLOS: Beyond Visual Line of Sight
C-UAS: Counter Unmanned Aerial Systems
CAR: Conflict Armament Research
COTS: Commercial-Off-The-Shelf
CT: Computerized Tomography
DARPA: Defense Advanced Research Projects Agency
DIB: Drone-In-A-Box
DJI: Da-Jiang Innovations
DOD: Department of Defense
DPRK: Democratic People's Republic of Korea
EMP: Electromagnetic Pulse
eVTOL: Electric Vertical Take-Off and Landing
EW: Electronic Warfare
FAA: Federal Aviation Administration
FCAS: Future Combat Air System
FSB: Federal Security Service
GAO: Government Accountability Office
GARC: Greenough Advanced Rescue Craft
GDP: Global Domestic Product

GPS: Global Positioning System
HAZMAT: Hazardous Material
HIMARS: High Mobility Artillery Rocket System
IADS: Integrated Air Defense System
IAMD: Integrated Air and Missile Defense
IDF: Israeli Defense Force
IED: Improvised Explosive Device
IRST: Infrared Search and Track
ISIS: Islamic State of Iraq and the Levant
ISR: Intelligence, Surveillance, and Reconnaissance
LiDAR: Light Detection and Ranging
LOCUST: Low-Cost Unmanned Aerial System Swarming Technology
LRUSV: Long-Range Unmanned Surface Vessel
MALD: Miniature Air Launched Decoy
MALE: Medium-Altitude Long Endurance
MCM: Mine Counter Measures
MRI: Magnetic Imaging Resonance
MRTT: Multi-Role Tanker Transport
MUM-T: Manned-Unmanned Teaming
NASA: National Aeronautics and Space Administration
NATO: North Atlantic Treaty Organization
NGAD: Next Generation Air Dominance
NCR-IADS: The National Capital Region-Integrated Air Defense System
NLAW: Next Generation Light Anti-Tank Weapon
NORAD: North American Aerospace Defense Command
OER: Oxygen Evolution Reaction
OSCE: Organization for Security and Cooperation in Europe
PLO: Palestine Liberation Organization
PPDS: Precision Payload Delivery System
PRC: People's Republic of China
RCV: Robotic Combat Vehicle
REPMUS: Robotic Experimentation and Prototyping with Maritime Uncrewed Systems
RF: Radio Frequency

ROK: Republic of Korea
RPG: Rocket Propelled Grenade
RUSI: Royal United Services Institute
SHORAD: Short Range Air Defense
SLAM-F (French): Système de lutte anti-mines futur ("Future Mine
    Warfare System")
sUAS: Small Unmanned Aerial System
TALONS: Towed Airborne Lift of Naval Systems
TCG (Turkish): Türkiye Cumhuriyeti Gemisi ("Republic of Turkey
    Ship")
THOR: Tactical High-power Operational Responder
TRX: Tracked Robot Demonstrator
TTPs: Tactics, Techniques, and Procedures
UAM: Urban Air Mobility
UAS: Unmanned Aerial Systems
UK: United Kingdom
UON: Urgent Operation Needs
UOSS: Unmanned Outer Space Systems
USAF: United States Air Force
USMC: United States Marine Corps
USS: Unmanned Surface Systems
UUS: Unmanned Undersea Systems
VTOL: Vertical Take-Off and Landing
XLUUV: Extra-Large Unmanned Undersea Vehicle

References

Aero Corner Editorial Team. n.d. "How to Legally Take Down a Drone." https://aerocorner.com/blog/legally-take-down-drone/.

Aitken, P. 2023a. "Researchers use AI to predict crops in Africa to help address food crisis." https://www.foxnews.com/world/researchers-use-ai-predict-crops-africa-help-address-food-crisis.

Aitken, P. 2023b. "Israel's new multimillion-dollar AI tank provides total battlefield vision: 'A new era.'" https://www.foxnews.com/world/israels-new-multimillion-dollar-ai-tank-provides-total-battlefield-vision-an-extraordinary-leap.

Aitken, F. 2023c. "German military plows millions into AI 'environment' for weapons tests that could change combat forever." https://www.foxnews.com/world/german-military-plows-millions-ai-environment-weapons-tests-change-combat.

Aitken, F. 2023d. "FBI chief warns that terrorists can unleash AI in terrifying new ways." https://www.foxnews.com/us/fbi-chief-warns-terrorists-can-unleash-ai-terrifying-new-ways.

Allen, G. 2019. "Understanding China's AI Strategy: Clues to Chinese Strategic Thinking on Artificial Intelligence and National Security." https://www.cnas.org/publications/reports/understanding-chinas-ai-strategy.

Allison, G. 2023. "British drone releases torpedo at sea for first time." https://www.ukdefencejournal. org.uk/british-drone-releases-torpedo-at-sea-for-first-time/.

Astute Analytica India Pvt. Ltd. 2022. "Unmanned Aerial Vehicle (UAV) Market to Reach US\$ 106.03 Billion by 2030." https://www.finance.yahoo.com/news/unmanned-aerial-vehicle-uav-market-133000966.html#amp_tf=From%20

%251%24s&aoh=16690443068994&csi=0&referrer=h
ttps%3A%2F%2Fwww.google.com.

Axe, D. 2022. "Russia's Electronic-Warfare Troops Knocked Out 90 Percent Of Ukraine's Drones." https://www.msn.com/en-us/news/world/russia-s-electronic-warfare-troops-knocked-out-90-percent-of-ukraine-s-drones/ar-AA15DrRb?ocid=msedgntp&cvid=204b0897ddbc49c4a955f0e8db82b0f3.

Bajak, F., and H. Arhirova. 2023. "Drone advances in Ukraine could bring dawn of killer robots." https://apnews.com/article/russia-ukraine-war-drone-advances-6591dc69a4bf2081dcdd265e1c986203.

Bajak, F. 2023. "Pentagon's AI initiatives accelerate hard decisions on lethal autonomous weapons." https://apnews.com/article/us-military-ai-projects-0773b4937801e7a0573f44b57a9a5942?.

Baker, S. 2023. "Russia's Iranian-made drones are powered by German technology stolen 17 years ago, experts say." https://www.msn.com/en-us/news/world/russia-s-iranian-made-drones-are-powered-by-german-technology-stolen-17-years-ago-experts-say/ar-AA1au44r?ocid=hpmsn&cvid=197d7aa81580438bad-2994106441c840&ei=34.

Bardsley, D. 2023. "Artificial intelligence: Is it getting out of control?" https://www.thenationalnews.com/world/2023/04/04/in-a-world-with-deepfake-videos-and-images-can-we-tell-reality-and-fiction-apart/?_gl=1*1r2z670*_ga*Tkl1Rl8waTNiRz-ZuSFFRSEFiUTVVZGlpQkZINDVhNFl6Wm1mRDUxam-1HaFhKTVFlWFg1dVphLXZBYTctMjB2cg.

Barnes, J. 2023. "How homemade drones are tilting the war in Ukraine's favour."https://www.yahoo.com/news/dispatch-homemade-drones-tilting-war-152235713.html

Beighton, R. 2021. "World's first crewless, zero emissions cargo ship will set sail in Norway." https://www.cnn.com/2021/08/25/world/yara-birkeland-norway-crewless-container-ship-spc-intl/index.html.

Bertrand, N. 2023. "CNN Exclusive: A single Iranian attack drone found to contain parts from more than a dozen US companies."

https://www.cnn.com/2023/01/04/politics/iranian-drone-parts-13-us-companies-ukraine-russia/index.html.

Bisht, I. 2023a. "Turkey Unveils Indigenous Kamikaze Drone." https://www.thedefensepost.com/2023/03/24/turkey-indigenous-kamikaze-drone-2/.

Bisht, I. 2023b. "Iran Unveils Long-Range Drone Armed with Air-to-Air Missile." https://www.thedefensepost.com/2023/10/27/iran-air-air-missile/?expand_article=1.

Bogaisky, J. 2023. "Billionaire Palmer Luckey Unveils A New Jet-Powered Drone—And It Looks Bonkers." https://www.msn.com/en-us/travel/news/billionaire-palmer-luckey-unveils-a-new-jet-powered-drone-and-it-looks-bonkers/ar-AA-1kOTbq?cvid=8b4323ef712d49868f158ca6ba105be4&ocid=winp2fptaskbar&ei=23&sc=shoreline.

Borrell, J. 2022. "The war in Ukraine and its implications for the EU." https://www.eeas.europa. eu/eeas/war-ukraine-and-its-implications-eu_en.

Bowden, M. 2022. "The Tiny and Nightmarishly Efficient Future of Drone Warfare." https://www.theatlantic.com/technology/archive/2022/11/Russia-ukraine-war-drones-future-of-warfare/672241/.

Boyes, R. 2022. "Drones over Ukraine are reinventing war." https://www.thetimes.co.uk/article/996ba4f8-6cde-11ed-b8ae-c57034d-fa905?shareToken=05019ea2612e722da7e08205 b5ef5cee.

Brimelow, B. 2023. "Shunned by the US, Turkey is preparing to launch its first aircraft carrier—but it will come with a twist. https://www.businessinsider.com/turkey-preparing-for-first-aircraft-carrier-tcg-anadolu-with-drones-2023-2.

Bronk, J., N. Reynolds, and J. Watling. 2022. "The Russian Air War and Ukrainian Requirements for Air Defense." https://www.rusi.org/explore-our-research/publications/special-resources/russian-air-war-and-ukrainian-requirements-air-defense.

Bryen, S. 2022. "South Korea's tech no match for intruding NK drones." https://asiatimes.com/2022/12/south-koreas-tech-no-match-for-intruding-nk-drones/.

Bush, D. 2023. "EXCLUSIVE: Russian Military Uses China in Sourcing Banned Tech from 59 U.S. Firms." www.newsweek.com/exclusive-russias-vast-sanctions-evasion-secures-us-european-tech-weapons-1807939.

Carlin, M. 2023. "The Air Force Wants 1,000 New Drones and a New Stealth Fighter." https://www.msn.com/en-us/news/world/the-air-force-wants-1-000-new-drones-and-a-new-stealth-fighter/ar-AA19680i?ocid=winp2fptaskbarent&cvid=6ffd80f-2c9e141d9b268 894c2328613f&ei=9.

Chapman, L., and E. Ludlow. 2023. "Palantir CEO: AI So Powerful 'I'm Not Sure We Should Even Sell This.'" https://finance.yahoo.com/news/palantir-ceo-touts-power-ai-171047233.html.

Childs, N. 2023. "Ukraine is ushering in a new uncrewed era at sea." https://www.defensenews.com/opinion/2023/09/11/ukraine-is-ushering-in-a-new-uncrewed-era-at-sea/.

ChinaLawTranslate.2017."ThePeople'sRepublicofChinaPRCNational Intelligence Law (as amended in 2018)." https://www.chinalawtranslate.com/en/national-intelligence-law-of-the-p-r-c-2017/.

Chopra, A. 2020. "Artificial Intelligence in Military Aviation." https://airpowerasia.com/2020/04/27/artificial-intelligence-in-military-aviation/.

Choudhury, M., A. Aoun, D. Badawy, L. Bacardit, Y. Marjane, and A. Wilkinson. 2021. "Letter dated 8 March 2021 from the Panel of Experts on Libya established pursuant to resolution 1973 (2011) addressed to the President of the Security Council *(2021)*." https://documents-dds-ny.un.org/doc/UNDOC/GEN/N21/037/72/PDF/N2103772.pdf?Open Element.

CIA. 2021. "Roadways." https://www.cia.gov/the-world-factbook/field/roadways/.

Cornish, N. 1998. "The Lagrange Points. WMAP Education and Outreach.

Culturalist Press. 2023. "What is a Black Hornet drone?" https://www.culturalistpress.com/black-hornet/.

Daleo, J. 2023. "Single Canadian Prison Sees Nearly 100 Illegal Drone Deliveries in 2022." https://www.flyingmag.com/single-canadian-prison-sees-nearly-100-illegal-drone-deliveries-in-2022/.

Davis, C. 2022. "Ukraine is getting more THeMIS robots, ground vehicles that Russian troops have been offered $16,000 to capture." https://www.businessinsider.com/ukraine-to-receive-themis-robots-russian-troops-encouraged-to-capture-2022-11.

Davis, J. 2022. "Report: 9,000 Drone Incursions into US Airspace Recorded; Reconnaissance Conducted on American Forces." https://www.westernjournal.com/report-9000-drone-incursions-us-airspace-recorded-reconnaissance-conducted-american-forces/.

Dawes, J. 2023. "War in Ukraine accelerates global drive toward killer robots." https://www.yahoo.com/news/war-ukraine-accelerates-global-drive-132417153.html.

Debusmann, B. 2022. "US shootings: Firm unveils plans for Taser-armed drones." https://www.bbc.com/news/world-us-canada-61685117?utm_campaign=mb&utm_medium=newsletter&utm_source=morning_brew.

Doll, S. 2022. "Autonomous cargo ship completes 500-mile voyage, avoiding hundreds of collisions." https://electrek.co/2022/05/13/autonomous-cargo-ship-completes-500-mile-voyage-avoiding-hundreds-of-collisions/.

Defense Security Cooperation Agency. n.d. "Qatar: PATRIOT Missile System and Related Support and Equipment." https://www.dsca.mil/press-media/major-arms-sales/qatar-patriot-missile-system-and-related-support-and-equipment.

Dunhill, J. 2021. "First 'AI-War': Israel Used World's First AI-Guided Swarm Of Combat Drones In Gaza Attacks." www.iflscience.com/first-ai-war-israel-used-worlds-first-aiguided-swarm-of-combat-drones-in-gaza-attacks-60221.

Dunne, P. 2013. "Miniature surveillance helicopters help protect front line troops." https://www.gov.uk/government/news/miniature-surveillance-helicopters-help-protect-front-line-troops.

Eastwood, B. 2022. "NGAD: New US Air Force Stealth Fighter Could Be 'Head Coach' For Drones." https://www.19fortyfive.com/2022/02/ngad-new-us-air-force-stealth-fighter-could-be-head-coach-for-drones/.

Eberhart, C. 2023. "Who is watching you? AI can stalk unsuspecting victims with 'ease and precision': experts." https://www.foxnews.com/us/who-is-watching-you-ai-can-stalk-unsuspecting-victims-ease-precision-experts.

Eckstein, M. 2022. "What's ahead for Navy unmanned underwater vehicle programs?" https://www.defensenews.com/naval/2022/11/29/whats-ahead-for-navy-unmanned-underwater-vehicle-programs/.

Edwards, B. 2023a. "Ukraine Is the World's Drone Laboratory: How Does This Transcend Security in Society?" https://www.securityindustry.org/2023/08/25/ukraine-is-the-worlds-drone-laboratory-how-does-this-transcend-security-in-society/.

Edwards, B. 2023b. "Mother plucker: Steel fingers guided by AI pluck weeds rapidly and autonomously." https://arstechnica.com/information-technology/2023/11/mother-plucker-steel-fingers-guided-by-ai-pluck-weeds-rapidly-and-autonomously/.

Federal Aviation Administration. n.d. "Unmanned Aircraft Systems (UAS)." https://www.faa.gov/uas/.

Getahun, H. 2022. "After an AI bot wrote a scientific paper on itself, the researcher behind the experiment says she hopes she didn't open a 'Pandora's box.'" https://www.businessinsider.in/international/news/after-an-ai-bot-wrote-a-scientific-paper-on-itself-the-researcher-behind-the-experiment-says-she-hopes-she-didnt-open-a-pandoras-box/articleshow/92771676.cms.

Gibson, A., A. Merchant, and B. Vigneron. 2020. "Autonomous Systems in the Combat Environment: The Key or the Curse to the U.S." https://thestrategybridge.org/the-bridge/2020/10/8/autonomous-systems-in-the-combat-environment-the-key-or-the-curse-to-the-us.

Goldfarb, A., and J. Lindsay. 2022. "Artificial Intelligence and the Human Context of War." https://nationalinterest.org/blog/

techland-when-great-power-competition-meets-digital-world/
artificial-intelligence-and-human.

Goldman, E. 2010. "Revolutions in Warfare." https://doi.
org/10.1093/acrefore/9780190846626.013.289.

Grey, S., M. Tamman, and M. Zholobova. 2022. "The global supply trail that leads to Russia's killer drones." https://www.yahoo.
com/news/tucker-laughs-nervously-stone-faced-043546911.
html.

Gritten, D. 2023. "Rare Israeli drone strike kills West Bank militants."
https://www.msn.com/en-us/news/world/rare-israeli-drone-strike-kills-
west-bank-militants/ar-AA1cSW8H?cvid=be3a85a966eb49d-
29b6ebd7d7d49d38b&ocid=winp2 fptaskbarhover&ei=4.

Guggina, D. 2022. "We're Bringing the Convenience of Drone
Delivery to 4 Million U.S. Households in Partnership with
DroneUp."https://corporate.walmart.com/newsroom/2022/05/24/
were-bringing-the-convenience-of-drone-delivery-to-4-mil-
lion-u-s-households-in-partnership-with-droneup.

Hagenenaars, J. 2020. "TU Delft maritime hydrogen drone flies longer
and greener." https://interestingengineering.com/innovation/
this-green-hydrogen-vtol-drone-can-fly-for-35-hours?utm_
source=Facebook&utm_medium=content&utm_cam-
paign=organic& utm_content=Apr27.

Hambling, D. 2023a. "Jam Buster: How Ukraine's 'Secret Weapon'
Shrugs Off Russian Radio Interference." https://www.popu-
larmechanics.com/military/a42922481/tricopter-drone-atlas-
pro-resists-russian-jamming/.

Hambling, D. 2023b. "Ukrainian Heavy Bomber Drones Drop
Anti-Tank Mines." https://www.forbes.com/sites/davidham-
bling/2023/10/04/ukrainian-heavy-bomber-drones-drop-anti-
tank-mines/?sh=359a34445c28.

Hambling, D. 2023c. "The U.S. Navy's Billion-Dollar Mystery
'Kamikaze Drones.'" https:www.19fortyfive.com/2023/05/the-
u-s-navys-billion-dollar-mystery-kamikaze-drones/.

Hambling, D. 2023d. "Ukraine's AI Drones Seek And Attack Russian
Forces Without Human Oversight." https://www.forbes.com/

sites/davidhambling/2023/10/17/ukraines-ai-drones-seek-and-attack-russian-forces-without-human-oversight/.

Hambling, D. 2022. "How Ukraine Perfected The Small Anti-Tank Drone." https://www.-forbes-com.cdn.ampproject.org/c/s/www.forbes.com/sites/davidhambling/2022/06/01/how-ukraine-perfected-the-small-anti-tank-drone/amp/.

Hammes, T. 2023. "Game-changers: Implications of the Russo-Ukraine war for the future of ground warfare." atlanticcouncil.org/in-depth-research-reports/issue-brief/game-changers-implications-of-the-russo-ukraine-war-for-the-future-of-ground-warfare/.

Harding, T. 2023. "AI forcing military to reimagine warfare." https://www.-thenationalnews-om.cdn.ampproject.org/c/s/www.thenationalnews.com/world/uk-news/2023/05/17/ai-forcing-military-to-reimagine-warfare/?outputType=amp.

Helfrich, E., and T. Rogoway. 2023. "Airbus's Drone Aerial Refueling Demo Is A Glimpse Of Air Combat Future." thedrive.com/the-war-zone/airbuss-autonomous-drone-refueling-demo-echoes-need-thats-set-to-rapidly-expand.

Hill, J. 2023. "Intelligent installation protects Middle East infrastructure." https://www.army-technology.com/news/marss-intelligent-installation-completed/.

Hohmann, W. 1925. *Die Erreichbarkeit der Himmelskörper* (*The Attainability of Celestial Bodies*). NASA Technical Translation F-44, 1960). https://archive.org/details/nasa_techdoc_19980230631.

Hu, C. 2022. "A tiny crabby robot gets its scuttling orders from lasers." https://www.popsci.com/technology/crab-shaped-micro-robots/.

Hunder, M. 2023. "Ground vehicles are the new frontier in Ukraine's drone war." https://www.reuters.com/world/europe/ground-vehicles-are-new-frontier-ukraines-drone-war-2023-07-13/.

Hurst, J. 2011. *Nathan Bedford Forrest: A Biography*. Knopf Doubleday Publishing Group, p. 247. ISBN 978-0-307-78914-3.

Hyeon-hwan, J., and K. Tae-gyu. 2023. "Korean defense firm to test unmanned vehicle for U.S. Marines." https://www.msn.com/en-us/news/world/korean-defense-firm-to-test-un-

manned-vehicle-for-u-s-marines/ar-AA1gIJSI?cvid=2b828988
42c14d70b4dada0707b75236&ocid=winp2fptaskbar&ei=19.

Hyundai Motor Group. 2021. "Genesis Celebrates Official Launch in China, Unveiling its All-New Vision of Automotive Luxury to Chinese Customers." https://www.hyundaimotorgroup. com/news/CONT0000000000001564.

Jacobsen, M. 2022. "The Dubious Prospects for Cargo Delivery Drones in Ukraine." https://warontherocks.com/2022/05/ the-dubious-prospects-for-cargo-delivery-drones-in-ukraine/.

Jahner, K. 2018. "Army wants mini-drones for its squads by 2018." https://www.armytimes.com/news/your-army/2016/04/03/ army-wants-mini-drones-for-its-squads-by-2018/.

Jain, A. 2021. "Did Turkish Combat Drones 'Autonomously' Attack Soldiers During The Libyan Civil War?" https://eurasiantimes. com/did-turkish-combat-drones-autonomously-attacked-soldi- ers-during-the-libyan-civil-war/.

Jankowicz, M. 2023. "Iranian-made drones cost as little as $20,000 to make but up to $500,000 to shoot down, a growing concern in Ukraine, report says." https://www.businessinsider.com/sui- cide-drones-much-cheaper-launch-than-shoot-down-ukraine- nyt-2023-1.

Jones, C. 2023. "Oshkosh Defense announces Robotic Combat Vehicle submission." https://defence-blog.com/oshkosh-de- fense-announces-robotic-combat-vehicle-submission/.

JPL. 2023. "Autonomous Systems Help NASA's Perseverance Do More Science on Mars." https:www.jpl.nasa.gov/news/auton- omous-systems-help-nasas-perseverance-do-more-science-on- mars.

Kaczmarek, S. n.d.a. "Robotic Swarm Intelligence for Lunar Exploration." https://sylvesterkaczmarek.com/blog/robotic-sw arm-intelligence-for-lunar-exploration/.

Kaczmarek, S. n.d.b. "Lunar Economy: Emerging Business Opportunities in Space." https://www.sylvesterkaczmarek.com/ blog/lunar-economy-emerging-business-opportunities-in-sp ace/.

Keller, J. 2023. "The Air Force's new directed energy weapon is ready to blast drone swarms out of the sky." https://www.msn.com/en-us/news/world/the-air-force-s-new-directed-energy-weapon-is-ready-to-blast-drone-swarms-out-of-the-sky/ar-AA1bq3vE?ocid=msedgntp&cvid=bf0e5150feb147b0a25f-81355cc4d289&ei=4.

Knight, W. 2022. "John Deere's Self-Driving Tractor Stirs Debate on AI in Farming." https://www.wired.com/story/john-deere-self-driving-tractor-stirs-debate-ai-farming/.

Knox, M., and W. Murray. 2001. *The Dynamics of Military Revolution, 1300–2050*. Cambridge: Cambridge University Press. ISBN 978-0-521-80079-2.

Knutsson, K. 2023. "Creepy Chinese drone swims underwater and flies through air." https:www/foxnews.com/tech/creepy-chinese-drone-swims-underwater-flies-air.

Kokkinidis, T. 2022. "Greek Inventor Creates Anti-drone System Called Minotaur." https://www.greekreporter.com/2022/11/29/greek-inventor-creates-anti-drone-system/.

Konaev, M., and S. Bendett. 2019. "Russian AI-Enabled Combat: Coming to a City near You." https://warontherocks.com/2019/07/russian-ai-enabled-combat-coming-to-a-city-near-you/.

Lee, M. 2023. "Robot chemist could create oxygen needed for colonizing Mars: study." https://www.foxnews.com/us/robot-chemist-could-create-oxygen-needed-colonizing-mars-study.

Levin, T. 2023. "A Toyota EV drove 1,200 miles without stopping to charge thanks to electric roads with wireless charging." https://www.businessinsider.com/toyota-ev-drives-1200-miles-range-wireless-charging-roads-electreon-2023-6.

Lewis, D. 2023. "Build This, Not That." https://www.usni.org/magazines/proceedings/2023/october/build-not.

Liebermann, O., H. Britzky, and N. Bertrand. 2023. "Russian fighter jet forces down US drone over Black Sea." hhps://www.cnn.com/2023/03/14/politics/us-drone-russian-jet-black-sea/index.html.

Losey, S. 2022. "How autonomous wingmen will help fighter pilots in the next war." https://www.c4isrnet.com/air/2022/02/13/how-autonomous-wingmen-will-help-fighter-pilots-in-the-next-war/.

Losey, S. 2023. "Air Force mulls remote control of drone wingmen." https://www.defensenews.com/air/2023/02/13/air-force-mulls-remote-control-of-drone-wingmen/.

Maci, V. 2022. "French military tees up new tech in rush to conquer the seabed." https://link.defensenews.com/click/26714125.99804/aHR0cHM6Ly93d3cuZGVmZW5zZW5ld3MuY29tL2ds-b2JhbC9ldXJvcGUvMjAyMi8wMi8xNC9mcmVuY2gt-bWlsaXRhcnktdGVlcy11cC1uZXctdGVjaC1p-bi1ydXNoLXRvLWNvbnF1ZXItdGhlL-XNlYWJlZC8/6127a2427c5f9829851e6049B4d88b846.

Makichuk, D. 2023. "Spectre of deadly drone attacks rears its head at Lula inauguration." https://davemakichuk.substack.com/p/danger-of-deadly-drone-attacks-rears.

McFadden, C. 2023a. "Ukraine is using cardboard drones to do battle with Russia now." https://interestingengineering.com/innovation/australia-ukraine-cardboard-drones.

McFadden, C. 2023b. "General Atomics' new drone radar can track balsa wood drones." https://interestingengineering.com/military/general-atomics-new-drone-radar.

McFadden, C. 2022. "Unmanned dogfight: Two drones have allegedly engaged in midair combat over Ukraine." https://interestingengineering.com/innovation/drone-on-drone-dogfight-ukraine.

McKinsey & Company. 2023. "The Economic Potential of Generative AI: The Next Productivity Frontier." https://www.mckinsey.com/capabilities/mckinsey-digital/our-insights/the-economic-potential-of-generative-AI-the-next-productivity-frontier#/.

McLaughlin, K. 2023. "Ukraine hosted a drone-building competition to see what new tech could make a difference in the war against Russia. msn.com/en-us/news/world/ukraine-hosted-a-drone-building-competition-to-see-what-new-tech-could-make-a-difference-in-the-war-against-russia/ar-AA1c5Ew8?ocid=h-pmsn&cvid=3642dd4c4cd84ee0b5c173bc2411cbcc&ei=14.

McMillan, T. 2023. "Pentagon Secretly Working to Unleash Massive Swarms of Autonomous Multi-Domain Drones to Dominate Enemy Defenses." https://thedebrief.org/pentagon-secretly-working-to-unleash-massive-swarms-of-autonomous-multi-domain-drones-to-dominate-enemy-defenses/.

McNeil, H. 2023. "Thales achieves breakthrough in Anglo-French maritime mine countermeasures." https://www.msn.com/en-us/money/companies/thales-achieves-breakthrough-in-anglo-french-maritime-mine-countermeasures/ar-AA1gExpp?cvid=470bd904b86c4db3817 e63189e0b95ca&ocid=winp-2fptaskbar&ei=38.

Menendez, B. 2017. "Putin: Leader in artificial intelligence will rule world." https://apnews.com/article bb5628f2a7424a10b3e38b-07f4eb90d4.

Mircea, C. 2022a. "Ehang Autonomous Aerial Vehicle Manufacturer Lands Its Largest Air Taxi Pre-Order in Japan." https://www.autoevolution.com/news/ehang-autonomous-aerial-vehicle-manufacturer-lands-its-largest-air-taxi-pre-order-in-japan-179736.html?utm_source=ae_self&utm_medium=ae_moreon&utm_campaign=ae_moreon_news_static.

Mircea, C. 2022b. "Natilus Announces $6 Billion in Purchase Commitments for Its Autonomous Cargo Aircraft." https://www.autoevolution.com/news/natilus-announces-6-billion-in-purchase-commitments-for-its-autonomous-cargo-aircraft-181354.html.

Mizokami, K. 2022." Russia desperately imported Iranian drones. Then, they malfunctioned in battle." https://www.yahoo.com/finance/news/russia-desperately-imported-iranian-drones-140700923.html.

Monks, K. 2023. "Clash of drones could decide the outcome of the Ukraine war, and shape future conflicts." https://inews.co.uk/news/world/clash-drones-ukraine-war-conflicts-2429906.

Moshtaghian, A., and M. Tawfeeq. 2023. "Israel was behind drone attacks at military plant in Iran, US media report." https://www.cnn.com/2023/01/30/middleeast/drone-strikes-Iran-isfahan-intl/index.html.

Muston, J. 2023. "Godfather of AI' says there's a 'serious danger' tech will get smarter than humans fairly soon." https://www.foxnews.com/tech/godfather-ai-serious-danger-tech-get-smarter-humans-fairly-soon.

NASA. n.d. "Jet Propulsion Laboratory." nasa.gov/centers/jpl/education/spaceprobe-20100225.html#:~:text=Sputnik%201%20was%20the%20first,by%20the%20former%20Soviet%20Union.

Nesnas, I., L. Fesq, and R. Volpe. 2021. "Autonomy for Space Robots: Past, Present, and Future." https://doi.org/10.1007/s43154-021-00057-2.

Newdick, T. 2023a. "Ukrainian Drones Target Moscow, Black Sea Fleet." https://www.thedrive.com/the-war-zone/ukrainian-drones-target-moscow-black-sea-fleet.

Newdick, T. 2023b. "Clandestine U.K. Program Developed 3D-Printed 'Suicide' Drone For Ukraine." https://www.thedrive.com/the-war-zone/clandestine-u-k-program-developed-3d-printed-suicide-drone-for-ukraine.

Newman, M. 2023. "Israel's new military AI systems select targets and plan missions 'in minutes.'" https://www.japantimes.co.jp/news/2023/07/17/world/israel-quietly-embeds-ai-military-systems/.

Niedercorn, F. 2023. "Fly-R launches its new R2-120 Raijin loitering munition." https://www.armyrecognition.com/defense_news_march_2023_global_security_army_industry/sofins_2023_fly-r_launches_its_new_r2-120_raijin_loitering_munition.amp.html.

Northwestern University. 2022. "Tiny robotic crab is smallest-ever remote-controlled walking robot." https://news.northwestern.edu/stories/2022/05/tiny-robotic-crab-is-smallest-ever-remote-controlled-walking-robot/.

O'Connor, T. 2017. "Russia's Military Challenges U.S. and China By Building A Missile That Makes Its Own Decisions." https://www.newsweek.com/russia-military-challenge-us-china-missile-own-decisions-639926.

Ohnsman, A. 2022. "Robotruck Startup Gatik Making Delivery Runs For Walmart Without Humans At The Wheel." https://www.forbes.com/sites/alanohnsman/2021/11/08/robotruck-startup-gatik-making-walmart-delivery-runs-without-human-drivers/?sh=233932440bff.

Oladimeji, S., and S. Kerner. 2022. "SolarWinds hack explained: Everything you need to know. https://www.techtarget.com/whatis/feature/SolarWinds-hack-explained-Everything-you-need-to-know.

Osborn, K. 2023a. "Stealthy New B-21 Bomber Will Fly Unmanned Missions & Control Drones." https://warriormaven.com/air/stealthy-new-b-21-bomber-will-fly-unmanned-missions-control-drones.

Osborn, K. 2023b. "NGAD Is Born: The Air Force's 6th Gen Stealth Fighter is Already Flying." https://www.msn.com/en-us/news/technology/ngad-is-born-the-air-force-s-6th-gen-stealth-fighter-is-already-flying/ar-AA18FnSX?ocid=winp2fptaskbarhover-ent&cvid=6517821e0bb747a4b21002bc36efc5d9&ei=45.

Osborn, K. 2023c. "Pentagon-Sent Drone Boats, Patrol Craft and Coastal Missiles Shut Down Russian Navy." https://warrior-maven.com/russia-ukraine/pentagon-sent-drone-boats-patrol-craft-and-coastal-missiles-shut-down-russian-navy.

Osborn, K. 2022. "Can Artificial Intelligence Help Kill Russian and Chinese Drone Swarms?" https://nationalinterest.org/blog/reboot/can-artificial-intelligence-help-kill-russian-and-chinese-drone-swarms-194621.

Osipova, N., and C. Thorbecke. 2022. "These fireball-dropping drones are on the frontlines of wildfire prevention." https://amp-cnn-om.cdn.ampproject.org/c/s/amp.cnn.com/cnn/2022/11/17/tech/drone-amplified/index.html.

Ozberk, T. 2023. "Ukraine's New Underwater Drone Marichka Breaks Cover." https://www.navalnews.com/naval-news/2023/08/ukraines-new-underwater-drone-marichka-breaks-cover/.

Panasovskyi, M. 2023a. "The makers of the Ukrainian kamikaze drone RAM II, which destroyed tens of millions of dollars worth of Russian equipment, could manufacture 1,000 of these UAVs each month." https://gagadget.com/en/230470-the-makers-of-the-ukrainian-kamikaze-drone-ram-ii-which-destroyed-tens-of-millions-of-dollars-worth-of-russian-equipment-/.

Panasovski, M. 2023b. "SARISA SRS-1X became the world's first quadcopter to launch a Hydra 70 missile. https://gagadget.com/en/243932-sarisa-srs-1x-became-the-worlds-first-quadcopter-to-launch-a-hydra-70-missile/.

Panasovsky, M. 2023c. "The world's first unmanned helicopter, the Golden Eagle, is unveiled and can deliver precision strikes from the air - the drone is equipped with a sniper rifle or assault rifle." https://gagadget.com/en/uav/249986-the-worlds-first-unmanned-helicopter-the-golden-eagle-is-unveiled-and-can-deliver-precision-strikes-from-the-air-the-/.

Panasovskyi, M. 2023d. "Swarm of drones controlled by artificial intelligence detected and tracked several targets on land and in the air." https://gagadget.com/en/weapons/252570-swarm-of-drones-controlled-by-artificial-intelligence-detected-and-tracked-several-targets-on-land-and-in-the-air/.

Panasovskyi, M. 2023e. "Natilus giant cargo drones will be powered by hydrogen engines with up to 2.5MW." https://gagadget.com/en/uav/252659-natilus-giant-cargo-drones-will-be-powered-by-hydrogen-engines-with-up-to-25mw/.

Panasovskyi, M. 2023f. "The Punisher drone with $1000 worth of ammunition destroyed $30m worth of Russian military equipment." https://gagadget.com/en/266054-the-punisher-drone-with-1000-worth-of-ammunition-destroyed-30m-worth-of-russian-military-equipment/.

Papadopoulos, L. 2023. "Germany to arm Ukraine with 5th generation tank-like drones to battle Russia." https://interestingengineering.com/innovation/tank-like-drones-to-battle-russia.

Papadopoulos, L. 2020. "This Green Hydrogen VTOL Drone Can Fly for 3.5 Hour." interestingengineering.com/innovation/this-green-hydrogen-vtol-drone-can-fly-for-35-hours?utm_source=Facebook&utm_medium=content&utm_campaign=organic&utm_content=Apr27.

Peng, Z. 2023. "Why China will win the global race for complete AI dominance." https://www.wired.co.uk/article/why-china-will-win-the-global-battle-for-ai-dominance.

Psaledakis, D., and A. Mohammed. 2023. "U.S. targets supply of Iranian drones to Russia in new sanctions." https://www.reuters.com/business/aerospace-defense/us-targets-supply-irani-an-drones-russia-new-sanctions-2023-01-06/#:~:text=WASH-INGTON%2C%20Jan%206%20(Reuters),during%20the%20conflict%20with%20Russia.

Raasch, J. 2023. "Next generation arms race could cause 'extinction' event akin to nuclear war, pandemic: tech chief." https://www.foxnews.com/tech/next-generation-arms-race-could-cause-ex-tinction-event-akin-nuclear-war-pandemic-tech-chief.

Rains, T. 2023. "United's Midnight air taxi is on track to debut by 2025—take a look at the plane. https://www.msn.com/en-us/travel/news/united-s-midnight-air-taxi-is-on-track-to-debut-by-2025-take-a-look-at-the-plane/ss-AA1c1H-b3?ocid=winp2fptaskbar&cvid=afd7d28dce184edbc3b0c-bcaf160629d&ei=23#image=1.

Rains, T. 2022. "Delta just gave the clearest indication yet of how electric flying taxis will change the experience of flying with airlines." https://www.businessinsider.com/delta-makes-a-60-million-investment-in-evtols-joby-aviation-2022-10.

Ramirez, V. 2022. "Drones as Big as 747s Will Fly Cargo Around the World With Low Emissions, Startup Says." https://singular-ityhub.com/2022/02/11/drones-as-big-as-a-747-will-fly-cargo-around-the-world-with-low-emissions/.

Reilly, D. 2023. "AI has power to 'manipulate' Americans, says Sen. Josh Hawley, advocates for right to sue tech companies." https:///www.foxnews.com/lifestyle/ai-power-manipulate-ameri-cans-sen-josh-hawley-advocates-right-sue-tech-companies.

Roque, A. 2023. "GDLS showcases short-range air defense pay-load on Tracked Robot 10-Ton." https://breakingdefense.com/2023/03/gdls-showcases-short-range-air-defense-payload-on-tracked-robot-10-ton/.

Rubin, U. 2023. "Russia's Iranian-Made UAVs: A Technical Profile." https://rusi.org/explore-our-research/publications/commentary/russias-iranian-made-uavs-technical-profile.

Saballa, J. 2023. "New Turkish Kamikaze Drone 'Replica' of Iranian Shahed-136." https://www.thedefensepost.com/2023/03/27/turkish-drone-replica-iran/.

Saphier, N. 2023. "AI is the future of health care, but here's what it can never replace: It's not just patients who may potentially benefit from AI in health care." https://www.foxnews.com/opinion/ai-future-health-care-heres-what-can-never-replace.

Sanborn, J. 2015. "Marines get a closer look at Black Hornet micro drone." https://www.marinecorpstimes.com/news/your-marine-corps/2015/09/22/marines-get-a-closer-look-at-black-hornet-micro-drone/.

Satam, P. 2023a. "China Unveils New Underwater=Launched Kamikaze Drone That Can Burst Open Taiwan's Defenses." https://eurasiantimes.com/chinas-unveils-new-underwater-launched-kamikaze-drone/?amp.

Satam, P. 2023b. "US Navy Hunts For 'Hunter Drones' That Can Operate From Non-Carrier Warships & Breach Enemy Defenses." https://www.eurasiantimes.com/us-hunts-for-hunter-drones-that-can-operate-from-non-carrier/?amp.

Schmidt, B. and Vance, A. 2020. "DJI won the drone wars, and now it's paying the price." https://www.bloomberg.com/news/features/2020-03-26/dji-s-drone-supremacy-comes-at-a-price?leadSource=uverify%20wall.

Searles, K. 2022. "Kawasaki trials cargo VTOL and robotic ground crew." https://www.roboticsandinnovation.co.uk/news/deliveries/kawasaki-trials-cargo-vtol-and-robotic-ground-crew.html.

Shayotovich, E. 2023. "This Futuristic Unmanned Tank Could Be A Game Changer For The US Army." https://www.msn.com/en-us/news/technology/this-futuristic-unmanned-tank-could-be-a-game-changer-for-the-us-army/ar-AA1gvk1I?ocid=hpmsn&cvid=44404595cb5143fbef8d151bff76233&ei=15.

Shestopalov, D. 2023. "Russia tests Sturm unmanned T-72 tank for deployment in Ukraine." https://life.ru/p/1606045.

Shoaib, A. 2023. "Ukraine shows off its next-generation 20-foot-long underwater drone that will sink Russian warships from below the surface." https://www.msn.com/en-us/news/world/ukraine-shows-off-its-next-generation-20-foot-long-underwater-drone-that-will-sink-russian-warships-from-below-the-surface/ar-AA1fQs9s?cvid=8e2c55a3ef84425cb51d8e597a97470a&ocid=winp2fptaskbarhover&ei=11.

Sicard, S. 2023. "This flame throwing robot dog proves we're all doomed." https://www.armytimes.com/off-duty/military-culture/2023/07/14/this-flame-throwing-robot-dog-proves-were-all-doomed/.

Skove, S. 2023. "Army moves ahead on Ukraine-style bomber drones." https://www.defenseone.com/technology/2023/10/army-moves-ahead-ukraine-style-bomber-drones/390918/.

Slotta, D. 2022. "Leading global drone manufacturers 2021, by share of sales volume." https://www.statista.com/statistics/1254982/global-market-share-of-drone-manufacturers/.

Smith, P. 2023. "North Korea tests underwater attack drone that can generate 'radioactive tsunami.'" https://www.nbcnews.com/news/world/kim-jong-un-north-korea-test-underwater-nuclear-attack-drone-rcna76470.

Spirlet, T. 2023. "Ukraine shows off a new ground drone meant to drive under Russian tanks and blow up." https://www.businessinsider.com/ukraine-ratel-drone-meant-to-drive-under-russian-tanks-explode-2023-10.

Study Finds. 2023. "Pentagon reportedly planning new 'weapon of mass destruction'—a horde of missile-carrying drones." https://studyfinds.org/weapon-of-mass-destruction-drones/.

Sutton, H. 2023a. "World's Largest Submarine Drone Being Built in Germany." https://www.navalnews.com/naval-news/2023/02/worlds-largest-submarine-drone-being-built-in-germany/.

Sutton, H. 2023b. "Ukraine's New Maritime Drone Is A Jet Ski." www.hisutton.com/Ukraine-Maritime-Drones-JetSki.html.

Sutton, H. 2022. "Ukraine's Maritime Drone Strikes Again: Reports Indicate Attack On Novorossiysk." https://www.navalnews.com/naval-news/2022/11/ukraine-maritime-drone-strikes-again-reports-indicate-attack-on-novorossiysk/.

Taylor, M. 2014. "Manned-Unmanned Teaming Training is the Key Ingredient." *U.S. Army Aviation* 2(1), 5. https://www.rucker.army.mil/aviationdigest/archive.html.

Tegler, E. 2022. "Bell Learned That Its APT 70 Cargo Drone Can Integrate With Manned Traffic During A Demo Flight. What Did It Learn About Avoidance and 5G?" https://www.forbes.com/sites/erictegler/2020/10/02/bell-learned-that-its-apt-70-cargo-drone-can-integrate-with-manned-traffic-during-a-demo-flight-what-did-it-learn-about-avoidance-and-5g/?sh=2cf5-b3622e1f.

Thunström, A. 2022. "We Asked GPT-3 to Write an Academic Paper about Itself—Then We Tried to Get It Published." https://www.scientificamerican.com/article/we-asked-gpt-3-to-write-an-academic-paper-about-itself-mdash-then-we-tried-to-get-it-published/.

Tiku, N. 2022. "The Google engineer who thinks the company's AI has come to life." https://www.washingtonpost.com/technology/2022/06/11/google-ai-lamda-blake-lemoine/.

Tiwari, S. 2022. "World's 1st Anti-Hypersonic System? China Says It Is Ready With An AI-Powered Defense Against Mach 5+ Missiles. https://eurasiantimes.com/worlds-1st-anti-hypersonic-system-china-says-its-is-ready/?amp.

Trevithick, J., and O. Parken. 2022. "China Launches Drone Ship That Acts As A Mothership For More Drones." https://sofrep.com/news/china-launches-ai-operated-mothership-zhu-hai-yun-for-drones-submersibles-and-boats/.

Trevithick, J., and T. Rogoway. 2022. "Russia's Black Sea Flagship Moskva Has Exploded Off Ukraine (Updated)." https://www.thedrive.com/the-war-zone/russias-black-sea-flagship-has-exploded-off-ukraine.

Trevithick, J. 2023a. "Stinger Missile-Toting Drone Boats Could Protect Navy Logistics Ships." https://www.thedrive.com/the-war-zone/stinger-missile-toting-drone-boats-could-protect-navy-logistics-ships.

Trevithick, J. 2023b. "A rocket launcher-toting robot dog could give Marines a valuable new way to remotely attack armored vehicles, especially in urban areas." https://www.the-drive.com/the-war-zone/marines-test-fire-robot-dog-armed-with-rocket-launcher.

Trevithick, J. 2023c. "Uncrewed submarines loaded with 'multi-domain' drones could offer a new dimension for hunting for enemy naval forces and mines, and more." https://www.thedrive.com/the-war-zone/drones-that-swim-and-fly-to-be-launched-recovered-by-uncrewed-submarine.

Trevithick, J. 2018. "Army Buys Small Suicide Drones To Break Up Hostile Swarms And Potentially More." https://www.thedrive.com/the-war-zone/22223/army-buys-small-suicide-drones-to-break-up-hostile-swarms-and-potentially-more.

Tripathi, A. 2023. "Ukraine War: Iran 'Modified' Shahed-136 Kamikaze Drones For Russia To Cause Max Destruction On Ukrainian Infra." https://world-defence.com/ukraine-war-iran-modified-shahed-136-kamikaze-drones-for-russia-to-cause-max-destruction-on-ukrainian-infra/.

Trock, D. 2023. "Skyborg: The Drone That Could Revolutionize Air Force Operations." https://www.msn.com/en-us/news/technology/skyborg-the-drone-that-could-revolutionize-air-force-operations/ar-AA1cL2bL?cvid=c99f5d9998a44e20977fd3500b-da8ab0&ocid=winp2fptaskbarhover&ei=12.

Turak, N. 2023. "Killer drones and multi-billion-dollar deals: Turkey's rapidly-growing defense industry is boosting its global clout." https://www.cnbc.com/2023/03/28/killer-drones-turkeys-growing-defense-industry-is-boosting-its-global-clout.html.

Tucker, P. 2023. "Is This New Microwave Weapon the Answer to Iranian Drones?" www.defenseone.com/technology/2023/01/new-microwave-weapon-answer-iranian-drones/382111/.

Tzu, S. n.d. AZQuotes.com. https://www.azquotes.com/author/19687-Sun_Tzu.

U.S. Air Force. 2019. "The U.S. Air Force Artificial Intelligence annex to the Department of Defense Artificial Intelligence Strategy." https://www.af.mil/Portals/1/documents/5/USAF-AI-Annex-to-DoD-AI-Strategy.pdf.

U.S. Army. 2017. "U.S. Army Robotics and Autonomous Systems Strategy." https://www.arcic.army.mil.

U.S. Congress. 2019. "H.R.5515-115th Congress: John S. McCain National Defense Authorization Act for Fiscal Year 2019." https://www.congress.gov/bill/115th-congress/house-bill/5515.

U.S. Government Accountability Office. 2022. "Weapon System Sustainment." https://www.gao.gov/assets/gao-23-106217.pdf.

U.S. Institute of Peace. 2023. "The Iran Primer." https://iranprimer.usip.org/blog/2023/mar/30/us-report-iran's-support-terrorism.

Vavasseur, X. 2022. "Here Is Our First Look At The US Navy's Orca XLUUV." https://www.navalnews.com/naval-news/2022/05/here-is-our-first-look-at-the-us-navys-orca-xluuv/.

Verger, R. 2021. "Skyborg's latest AI drone test is a preview of the future of air combat." https://www.popsci.com/technology/air-force-continues-skyborg-program/.

Vergun, D. 2023. "DOD Will Deploy AI-Enabled Detection System to Monitor D.C. Airspace." https://www.defense.gov/News/News-Stories/Article/Article/3507329/DOD-will-deploy-ai-enabled-detection-system-to-monitor-dc-airspace/.

Verma, P. 2022. "The military wants 'robot ships' to replace sailors in battle." https://www.washingtonpost.com/technology/2022/04/14/navy-robot-ships/.

Victor, D., and D. Kirkpatrick. (2019). "Iran Shoots Down a U.S. Drone, Escalating Tensions." *The New York Times*. ISSN 0362-4331.

Vigiarolo, B. 2023. "Feds collar suspected sanctions-busting Russian smugglers of US tech Parts sent to Moscow allegedly found on Ukrainian battlefields." https://www.msn.com/en-us/news/technology/feds-collar-suspected-sanctions-busting-russian-smugglers-of-us-tech/ar-AA1je6OQ?cvid=0edd347ab7284bb-4cf3e800eec96962f&ocid=winp2fptaskbar&ei=26.

Wang, B. 2023. "Replicator Program Will Scale Up to Thousands of Smart AI Combat Drones Within 24 Months." https://www.nextbigfuture.com/2023/09/replicator-program-will-scale-up-to-thousands-of-smart-ai-combat-drones-within-24-months.html.

Whalen, J. 2022. "Russian drones shot down over Ukraine were full of Western parts. Can the U.S. cut them off? https://www.washingtonpost.com/technology/2022/02/11/russian-military-drones-ukraine/.

Whitney, J. 2021. "Artificial intelligence and machine learning for unmanned vehicles." https://www.militaryaerospace.com/unmanned/article/14202040/artificial-intelligence-and-machine-learning-for-unmanned-vehicles.

Williams, D. 2022. "Bahrain hosts Bennett as Israel wades into Gulf security." https://www.reuters.com/world/europe/us-navy-eyes-israeli-robot-boats-bennett-visits-Bahrain-official-says-2022-02-15/?utm_source=Sailthru&utm_medium=email&utm_campaign=EBB%2002.15.2022&utm_term=Editorial%20-%20Early%20Bird%20Brief.

Whitney, J. 2021. "Artificial intelligence and machine learning for unmanned vehicles." https://www.militaryaerospace.com/unmanned/article/14202040/artificial-intelligence-and-machine-learning-for-unmanned-vehicles.

Wodecki, B. 2022. "Tiny AI Kamikaze Drone Packs a Punch." https://www.iotworldtoday.com/2022/11/28/tiny-ai-kamikaze-drone-packs-a-punch/.

Yaron, O. 2022. "Israel's Elbit Sytems Unveiled Micro-suicide Drone, and It Has a Mother Ship." https://haaretz.com/israel-news/security-aviation/2022-11-18/ty-article/.premium/israels-elbit-systems-unveiled-micro-suicide-drone-and-it-has-a-mother-ship/00000184-84d2-dd3b-a5bf-c4d6e2880000?v=1669219647915.

Zoldi, D. 2022. "This Company Just Unveiled The World's Largest Autonomous Commercial Drone Deployment." https://www.forbes.com/sites/dawnzoldi/2022/04/26/this-company-just-unveiled-the-worlds-largest-autonomous-commercial-drone-.deployment/?sh=56500c9f432c.

Zoldi, D. 2021. "Drone Swarms: the Good, the Bad, and the Beautiful." https://dronelife.com/2021/05/02/drone-swarms-the-good-the-bad-and-the-beautiful/.

Zwiezen, Z. 2023. "Uh-Oh: USAF's Killer AI Drone Sounds Straight Out Of Horizon Zero Dawn." https://www.msn.com/en-us/lifestyle/shopping/uh-oh-usaf-s-killer-ai-drone-sounds-straight-out-of-horizon-zero-dawn/ar-AA1c0O98?cvid=bcb70763fcfe44c6d5d953ab4592b70b&ocid=winp2fptaskbar&ei=5.

About the Author

Dr. Terence M. Dorn was born in Ankara, Turkey. As the son of an Air Force chief master sergeant, he traveled worldwide. He eventually attended college in Nebraska, earning two baccalaureate of art degrees in business administration and Sociology from Bellevue College, later reaccredited as a university.

Dr. Dorn joined the Army in 1985, excelled in the Army's Air Defense branch, and deployed seven times to various combat zones. Three noteworthy assignments included serving as a speechwriter to the chief of staff of the Army; military assistant to the secretary of defense; and deputy commanding officer of the International Security Assistance Force, Turkish-led Regional Command Capital in Kabul, Afghanistan. In 2013, he was inducted into the Army Officer Candidate School Hall of Fame and retired in 2014.

While serving the nation as an Army officer, Dr. Dorn earned a master of arts degree in international relations from Boston University and a master of science degree in national security strategy from the National War College. In 2020, he completed his PhD in business administration from Northcentral University. His area of specialization was in Homeland Security, and his dissertation was titled *A Phenomenological Study Examining the Vulnerabilities of U.S. Nuclear Power Plants to Attack by Unmanned Aerial Systems.*

Dr. Dorn has authored *Quotes: The Famous and Not So Famous, Unmanned Systems: Savior or Threat, U.S. Critical Infrastructure: Its Importance and Vulnerabilities to Cyber and Unmanned Systems*, and now *The Unmanned Systems and Artificial Intelligence Revolution.* In addition, Dr. Dorn wrote *Countering the Threat Posed by Unmanned Systems Armed with Weapons of Mass Destruction* while working at the Department of Homeland Security and its corresponding strategy implementation plan to turn that strategy into an executable reality.

Dr. Dorn's civilian honors include induction into the following honor societies: Academy of Criminal Justice Sciences, Alpha Phi Sigma, Iota Pi; the National Society of Leadership and Success, Sigma Alpha Pi; and the International Business Honor Society, Delta Mu Delta.

Dr. Dorn has spoken on the capabilities and threats posed by unmanned systems to US critical infrastructure to the following groups: National Geospatial-Intelligence Agency (NGA), VA; Transportation Security Administration (TSA), DC; Defense Threat Reduction Agency (DTRA), DC; Federal Bureau of Investigation (FBI), DC; Federal Aviation Administration (FAA), DC; US Secret Service (USSS), DC; InfraGard National Members Alliance (INMA), DC; NGA, DOD, Defense Intelligence Agency (DIA), MO; Government Coordinating Council (GCC), Nuclear Sector Coordinating Council (NSCC), Critical Infrastructure Partnership Advisory Council (CIPAC), DC; National Geospatial-Intelligence Agency (NGA), MO; and American Public University System, School of Security and Global Studies, DC.

Dr. Dorn is married to the former Nicole D. Holda, a retired major who served in the US Army's Military Intelligence Corps for over two decades.